ANNALS of THE NEW YORK ACADEMY OF SCIENCES

The New York Academy of Sciences
7 World Trade Center
250 Greenwich Street, 40th Floor
New York, NY 10007-2157

annals@nyas.org
www.nyas.org/annals

Published by Blackwell Publishing
On behalf of the New York Academy of Sciences

Boston, Massachusetts
2012

ANNALS *of* THE NEW YORK ACADEMY OF SCIENCES

VOLUME 1259

ISSUE

Environmental Stressors in Biology and Medicine

ISSUE EDITOR

Giuseppe Valacchi

University of Siena

TABLE OF CONTENTS

Ann. N.Y. Acad. Sci. ISSN 0077-8923

ANNALS OF THE NEW YORK ACADEMY OF SCIENCES
Issue: *Environmental Stressors in Biology and Medicine*

Omics approaches in cystic fibrosis research: a focus on oxylipin profiling in airway secretions

Jason P. Eiserich,[1,2] Jun Yang,[3] Brian M. Morrissey,[1] Bruce D. Hammock,[3] and Carroll E. Cross[1,2]

[1]Department of Internal Medicine, [2]Department of Physiology and Membrane Biology, [3]Department of Entomology, University of California, Davis, California

Address for correspondence: Carroll E. Cross, University of California, Davis School of Medicine, 4150 V Street, Suite 3400, Sacramento, CA 95817. cecross@ucdavis.edu

Cystic fibrosis (CF) is associated with abnormal lipid metabolism, intense respiratory tract (RT) infection, and inflammation, eventually resulting in lung tissue destruction and respiratory failure. The CF RT inflammatory milieu, as reflected by airway secretions, includes a complex array of inflammatory mediators, bacterial products, and host secretions. It is dominated by neutrophils and their proteolytic and oxidative products and includes a wide spectrum of bioactive lipids produced by both host and presumably microbial metabolic pathways. The fairly recent advent of "omics" technologies has greatly increased capabilities of further interrogating this easily obtainable RT compartment that represents the apical culture media of the underlying RT epithelial cells. This paper discusses issues related to the study of CF omics with a focus on the profiling of CF RT oxylipins. Challenges in their identification/quantitation in RT fluids, their pathways of origin, and their potential utility for understanding CF RT inflammatory and oxidative processes are highlighted. Finally, the utility of oxylipin metabolic profiling in directing optimal therapeutic approaches and determining the efficacy of various interventions is discussed.

Keywords: cystic fibrosis; inflammation; omics; oxylipins; respiratory tract secretions

Introduction

Cystic fibrosis (CF) is a genetic disease ascribed to mutations in the *CFTR* gene, which codes for the widely expressed CFTR chloride channel. Clinically, the disease manifestations include pancreatic insufficiency, intestinal malabsorption, and respiratory tract (RT) defects in fluid and electrolyte balance, mucociliary clearance, host defenses against microbes, and heightened inflammatory/immune responses. The resulting nutritional deficiencies include abnormalities in lipid absorption, including lipophilic antioxidant micronutrients, whereas the RT abnormalities are dominated by progressively more severe bacterial infection and the overly aggressive inflammatory response leading to lung destruction and premature death.[1]

Inflammation and oxidative stress in the CF RT

The interrelated RT pathobiology of CF involves chronic airway infection by microorganisms overcoming CF RT host defenses, and a dysregulated and heightened RT inflammatory response characterized by a progressively more intense sustained neutrophil recruitment and activation.[2] Abundant evidence supports the ongoing pro-oxidative environment of CF RT secretions,[3–5] which is largely attributed to activation of the NADPH oxidase (Nox-2) and further exacerbated by the catalytic activity of myeloperoxidase (MPO) of the recruited airway neutrophils, including contributions of pro-oxidant species released from microorganisms, such as *Pseudomonas aeruginosa*, and from endogenous RT

doi: 10.1111/j.1749-6632.2012.06580.x

epithelial cells (i.e., Duox enzymes). The presence of elevated levels of the nonenzymatic lipid peroxidation products of arachidonic acid (AA) (i.e., isoprostanes) in CF plasma, buccal mucosa cells, breath condensate, and bronchoalveolar lavage fluids attests to the pro-oxidant status of the CF RT. Although active RT oxidative processes represent a well-recognized hallmark in CF RT secretions affecting both host and resident microbe biology,[4,6] efficacious anti-inflammatory and antioxidant therapies in CF have not met with major success. For example, inhaled GSH trials have failed to improve any standard biomarkers of RT oxidative stress.[7,8] *N*-acetylcysteine (NAC) administration trials in CF patients have also failed to improve RT biomarkers of oxidative stress or yield convincing evidence of efficacy.[9] It can be concluded that, despite two decades of efforts, there is no strong case for supplemental antioxidant administrations beyond those to sustain normal plasma levels of the lipophilic antioxidants (e.g., vitamin E).

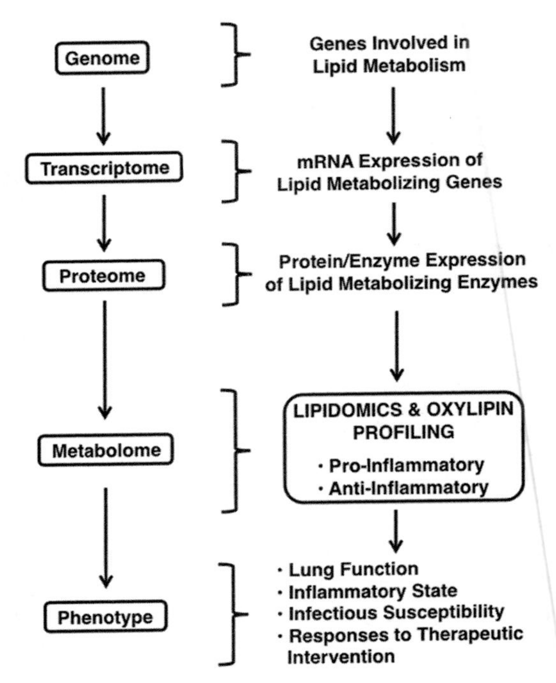

Figure 1. Generalized and lipid-specific scheme to illustrate omics pathways from gene to phenotype with regard to understanding CF airway biology and pathobiology.

Omic approaches in CF

Newer methods for more qualitative and quantitative characterization of the pathobiology driven by *CFTR* mutations giving rise to CF phenotypes are integrating with the myriad of molecular biology that has emerged in the two decades since discovery of the *CF* gene. These new-era technologies include those focused on systems-level approaches to global analysis of cell, tissue, and organ CF tissues.[10,11] As illustrated in Figure 1, these technologies focus on mutant CFTR effects on cellular message levels (genomics/transcriptomics), cellular protein and protein networks (proteomics), lipids and lipid metabolic pathways (lipidomics), and overall integrative omics (metabolomics). All of the new omics approaches are in the quest of further understanding CF pathophysiology that could reveal strategies for improved therapeutic approaches and clinical patient outcomes.

Many of the omics approaches have been targeted toward increased understanding of how mutated CFTR impacts on basic cellular functions beyond its primary role in chloride ion transport.[12–19] Other omics studies have focused on CF tissues,[18,20–22] CF blood compartments,[23–25] bronchoalveolar lavage fluid,[26,27] and exhaled breath condensate.[28,29] These latter two fluids have the advantage that they are collected directly from the RT, the tissue of greatest impact by CF. However, these two methods have the disadvantages that they are approximately 100 times and 10,000 times diluted with water, respectively, and neither collects secretions exclusively from the airways, the lung compartment most directly affected in CF. Several of the omic interrogations have incriminated the involvement of inflammatory-immune system genes in determining the severity of a given mutant *CFTR* genotype's phenotype.[30–32] Other omic studies have presented evidence suggesting a mechanistic link between mutations in CFTR and acquisition of the proinflammatory phenotype in the CF RT.[33] Mutant CFTR effects on CF neutrophil function may even represent manifestations of one such activity modulating neutrophil function,[34] although this may be related to the compromised chloride ion transport function associated with mutant CFTR.[35] A recent paper has presented evidence for an even more widespread mutant CFTR modulation of function in cells of the inflammatory/immune system that is generally recognized.[36]

The overall field of lipidomics has progressed somewhat slower than that of genomics and proteomics. Contributing factors could include the diversity of biomolecular lipid classes, including fatty acids and their derivatives (amides, esters, and oxygenated species), phospholipids, di- and triglycerides, sphingolipids, sterols, and numerous nonenzymatic oxylipids (e.g., isoprostanes). Lipid isolation from complex tissue matrices, extraction efficiencies, availability of standards, bioinformatic tools, and integration with imaging and other omics databases are only just being refined. Nonetheless, technologies are advancing rapidly.[37,38] Some of these studies have addressed oxidized phospholipids,[39] including those in CF patients.[40,41] Many papers have focused on lipidomic profiling of inflammatory lipid mediators (lipoxins) including the well-known eicosanoids synthesized from AA (20:4), as well as the major omega-3 polyunsaturated fatty acids eicosapentaenoic acid (EPA, 20:5) and docosohexaenoic acid (DHA, 22:6).[42–45] Two excellent recent reviews have detailed protocols and addressed lipid mediator profiling in various CF tissue matrices.[46,47]

Oxylipin profiling of CF RT airway secretions

The term *oxylipin* (sputum) covers a broad spectrum of compounds, many of which have high biological activities and which are formed from unsaturated fatty acids in a cascade of reactions, most of which include at least one step of mono- or dioxygen-catalyzed oxygenation. There are several reasons for focusing on RT oxylipins in CF, a disease known to be associated with abnormalities of lipid absorption, both in extracellular and cellular lipid constituencies.[48–56] AA levels appear higher, whereas longer chained 20:5 and 22:6 fatty acids appear reduced, probably largely (but not only) as a function of their reduced absorption and dysregulated metabolism.[57,58] Recently, genes in the AA–prostagladin–endoperoxide synthetase pathway have been reported as possible modulators of disease severity in CF patients.[59] A wide spectrum of oxidative lipid products, most notably those of 20:4, 20:5, and 22:6, are increasingly being recognized to be important regulators of the intensity and duration of acute and chronic inflammatory pathways,[43,45,59–64] including those initiated in the lung by P450 pathways upon microbial exposure.[65] Of relevance, there are several reports of abnormalities of polyunsaturated lipid oxidation products, such as the lipoxins in inflammatory RT CF secretions, that have been reported.[66,67]

As of 2011, we are not aware of any published studies applying broad-spectrum oxylipin profiling methods to investigate the intensely infected and inflamed CF RT. The resident pool of oxylipins can be expected to be embedded in a complex matrix of CF sputum[68,69] that itself has both pro-oxidant and antioxidant properties.[3] Although it is relatively easy to collect freshly expectorated RT secretions from CF patients, specimen processing including measures to prevent further artifactual oxidation, lipid extraction methods, and strategies for evaluating extraction efficiency and data expression present practical, yet surmountable challenges. In several cases, key validation compounds (quantitative internal standards) of interest are not readily available from individual or corporate investigators. This has particularly presented obstacles in the determination of several of the more recently described oxidized polyunsaturated fatty acids with novel

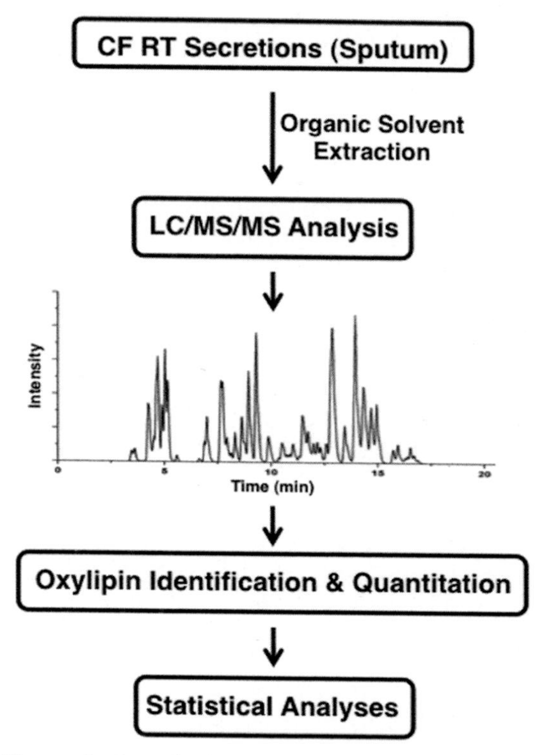

Figure 2. Generalized scheme illustrating an LC/MS/MS metabolomic approach to comprehensively evaluate bioactive oxylipin profiles in CF RT secretions.

anti-inflammatory properties. The CF mucus oxylipin profile can be expected to represent the omics of not only most inflammatory cells and RT epithelial cells but also reflect the contributions of a wide spectrum of airway microbiota[70] including *P. aeruginosa*, in the context of their residence in the CF airway.[71,72]

Despite these challenges, our laboratory groups have embarked upon developing methods (both sample preparation and analytical) to ascertain the profile of oxylipins in RT secretions from subjects capable of producing spontaneously expectorated sputum, as is the case for most adult CF patients. We have found that liquid–liquid solvent extraction of freshly obtained whole expectorated sputum is the most highly efficient method for recovering oxylipins. We preserve/stabilize the specimens with butylated hydroxyltoluene, triphenylphosphine, and a broad spectrum COX inhibitor (indomethacin). The specimens are immediately frozen on dry ice and we take all practical steps to minimize exposure to air (it would be preferable to overlay an inert gas) during the subsequent work-up procedures. We subsequently use a state-of-the-art LC/MS/MS analytical method previously optimized,[73] and which is capable of screening nearly 100 diverse oxylipins within a single LC run of 20 minutes (Fig. 2). Thus far, we have been capable of identifying greater than 30 oxylipins in adult CF sputums. As illustrated in Figure 2, employing this analytical procedure, and coupled with multiple statistical methods, we have established that this oxylipin metabolomic method provides a potentially unique view into the complex inflammatory-immune processes occurring in the CF airway. A general scheme that illustrates our method is depicted in Figure 2. Of the detected oxylipins, approximately 75% are derived from AA, 20% from linoleic acid (LA), and a much smaller quantity from EPA and DHA. The predominant source of oxylipin metabolites in CF sputum is the 12-lipoxygenase pathway (12-LOX; ∼50%), followed by 5-lipoxygenase (5-LOX; ∼35%), cyclooxygenase (∼5%), 15-LOX (∼3%), and the remainder coming from P-450 dependent pathways. Not surprisingly, the predominant oxylipin metabolites present in the CF RT are typically regarded as proinflammatory. Much lower levels of anti-inflammatory oxylipins were detected including Resolvin E1. Interestingly, Lipoxin A4 was not detected in any of the CF sputum specimens we examined,

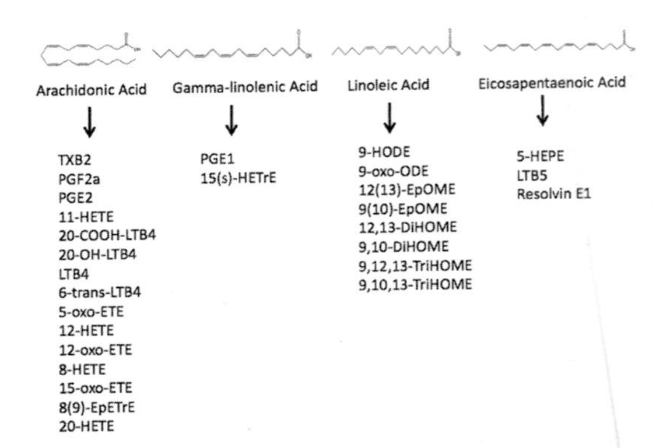

Figure 3. Summary of oxylipins detected in adult CF sputum as a function of the parent unsaturated fatty acid.

suggesting a deficiency of this anti-inflammatory lipid mediator. Figure 3 summarizes the oxylipins we have detected in CF sputum to date using our liquid chromatography/massspectrometry/mass spectrometry (LC/MS/MS) method, inclusive of their parent fatty acid. Preliminary analyses using the oxylipin metabolome in aggregate, as opposed to individual metabolite measurements, provides a more robust association with overall lung function (assessed by forced expiratory volume in 1 second [FEV1], % of predicted). The complete data and discussion of these findings are the topic of a recently accepted manuscript.[74]

Potential microbial contribution to oxylipins in CF sputum

Although the prevailing wisdom would indicate that oxylipins present in the CF airway likely primarily arise from host cells and tissues/lipid substrates, the presence of large numbers of bacteria (particularly *P. aeruginosa*) begs the question as to the possible involvement of microorganisms in the synthesis and metabolism of the analyzed RT oxylipins. *P. aeruginosa* contain primarily short chain saturated fatty acids,[75] thus they are not a source of the parent fatty acids of the oxylipins we have detected in our studies. However, recent studies have revealed that *P. aeruginosa* express a secreted cytotoxin (ExoU) with phospholipase activity capable of liberating free unsaturated fatty acids (LA and AA) from host cells.[76] Moreover, *P. aeruginosa* express a number of fatty acid metabolizing enzymes, including dioxygenases, hydroperoxide isomerases, and arachidonate 15-lipoxygenase,[77,78] that may directly contribute to the oxygenation of fatty acids in the CF airway.

A recent study has also identified an epoxide hydrolase produced by *P. aeruginosa*[79] that could potentially be used to convert epoxyeicosanoic acids (EETs), thus diminishing the anti-inflammatory functions of these oxylipins in the CF RT. The possibility that *P. aeruginosa*, and potentially other bacteria, may play a key role in the synthesis and metabolism of bioactive oxylipins in the CF airway is particularly exciting and remains a fertile area of investigation.

Translational implications

It is not difficult to envision how comprehensive oxylipin profiling could be incorporated into an increased understanding of the inflammatory versus anti-inflammatory bioactive lipid contributions to inflammatory-immune processes at RT apical cellular surfaces. Such mechanistic data could be incorporated into clinical trials focused on ameliorating the excessive RT inflammation in CF, and inform efficacies of systemic versus aerosolized treatments, when linked to appropriate patient outcome studies.[80–82]

RT oxylipin profiling in CF should provide dose-ranging insights with regard to nutritional and drug therapies targeted to modify bioactive lipid contributions to RT inflammatory pathobiology. These include the overly exuberant proinflammatory processes in CF that are possibly related to the high AA and low DHA/EPA concentrations observed in patients with CF, and the degree to which CF lipid and oxylipin abnormalities can be influenced by dietary[49,83–87] or pharmacological[88–90] interventions, including even antibiotics with anti-inflammatory activities that are used in treating CF patients,[91] statins,[92,93] and antiproteases.[94] It is important to note that some of these approaches are likely to modulate both host and microbe, and their interactions, in addition to solely affecting host inflammatory pathways. A theoretical list of emerging CF therapies and approaches that could influence CF RT lipidomics and reveal anti-inflammatory strategies are depicted in Table 1. Such approaches appear strengthened by recent data showing that select anti-inflammatory oxylipins (or their fatty acid precursors) appear to both facilitate antimicrobial activity and decrease inflammatory processes,[95] stimulating experimental activities designed for focal delivery of these compounds directly to the CF RT via inhalation routes.

Table 1. Therapeutic approaches that could influence CF RT oxylipin profiles

Modulators of cyclooxygenase (COX) and leukotriene pathways
Modulators of RT lipid metabolism
Statins
Phospholipase inhibitors
Cytochrome P450 modulators
PPAR gamma agonists
Soluble epoxide hydrolase inhibitors
Antibiotics with anti-inflammatory properties, such as azithromycin and tetracycline
Inhaled antiproteases
Sildenafil
Oral, systemic, or inhaled anti-inflammatory oxylipins

Clinical studies of the full inflammatory/anti-inflammatory profile of oxylipins in the CF RT are needed to more fully unveil significance of anti-inflammatory oxylipin-based therapies in the disease. Many of these oxylipins are already known to be therapeutic targets of and contributors of inflammatory-immune processes, but their efficacy within the RT airway fluids has not been extensively investigated. As therapeutic manipulation of both proinflammatory and anti-inflammatory oxylipin profiles are increasingly being proposed as effective RT disease modifiers, it can be speculated that more than one oxylipin target or pathway will be needed for addressing this aspect of anti-inflammatory therapy in CF.

Summary

The purpose of identifying CF omic signatures is to provide clinically useful clues for improved patient management and disease outcome. Further omic characterization of the infected, inflamed, and oxidizing RT fluids (separating the atmospheric environment from the underlying RT cells) should provide for new insights into CF disease pathobiology. Such omic approaches should in addition mechanistically inform new targets for anti-inflammatory and antioxidant therapies and complement clinical studies specifically designed to decrease overzealous RT inflammatory-immune pathways. Interestingly, eicosanoid lipidomics are starting to be examined in human breath condensates.[42] Finally, detailed studies of the inflamed CF RT secretions could lead to further understanding and therapeutics of the broad swath of RT diseases characterized by activation of RT inflammatory-immune processes, including those initiated by toxic inhaled environmental agents.

Acknowledgments

The authors would like to thank the patient volunteers who participated in the study. The work was supported by a fellowship from Cystic Fibrosis Research, Inc. (J.Y.), the Cystic Fibrosis Foundation support of the UCD Adult CF Program (CEC and BMM), the NIH (HL092506, J.P.E. and C.E.C.), and in part by NIEHS SBRP Grant p42 ES004699, NIEHS Grant R37 ES02710, and NIH/NIEHS Grant R01 ES013933 (B.D.H.).

Conflicts of interest

The authors declare no conflicts of interest.

References

1. Rowe, S.M, S. Miller & E.J. Sorscher. 2005. Cystic fibrosis. *N. Engl. J. Med.* **352:** 1992–2001.
2. Downey, D.G., S.C. Bell & J.S. Elborn. 2009. Neutrophils in cystic fibrosis. *Thorax* **64:** 81–88.
3. van der Vliet, A., J.P. Eiserich, G.P. Marelich, *et al.* 1997. Oxidative stress in cystic fibrosis: does it occur and does it matter? *Adv. Pharmacol.* **38:** 491–513.
4. Cantin, A.M., T.B. White, C.E. Cross, *et al.* 2007. Antioxidants in cystic fibrosis. Conclusions from the CF antioxidant workshop, Bethesda, Maryland, November 11–12, 2003. *Free Radic. Biol. Med.* **42:** 15–31.
5. Galli, F., A. Battistoni, R. Gambari, *et al.* 2012. Oxidative stress and antioxidant therapy in cystic fibrosis. *Biochim. Biophys. Acta* **1822:** 690–713.
6. Chang, W., D.A. Small, F. Toghrol & W.E. Bentley. 2005. Microarray analysis of Pseudomonas aeruginosa reveals induction of pyocin genes in response to hydrogen peroxide. *BMC Genomics* **6:** 1–14.
7. Griese, M., J. Ramakers, A. Krasselt, *et al.* 2004. Improvement of alveolar glutathione and lung function but not oxidative state in cystic fibrosis. *Am. J. Respir. Crit. Care Med.* **169:** 822–828.
8. Hartl, D., V. Starosta, K. Maier, *et al.* 2005. Inhaled glutathione decreases PGE2 and increases lymphocytes in cystic fibrosis lungs. *Free Radic. Biol. Med.* **39**: 463–472.
9. Duijvestijn, Y.C. & P.L. Brand. 1999. Systematic review of N-acetylcysteine in cystic fibrosis. *Acta Paediatr.* **88:** 38–41.
10. Balch, W.E. 2011. Introduction to section II: omics in the biology of cystic fibrosis. *Methods Mol. Biol.* **742**: 189–191.
11. Drumm, M.L, A.G. Ziady & P.B. Davis. 2012. Genetic variation and clinical heterogeneity in cystic fibrosis. *Annu. Rev. Pathol.* **7**: 267–282.
12. Srivastava, M., O. Eidelman & H.B. Pollard. 1999. Pharmacogenomics of the cystic fibrosis transmembrane conductance regulator (CFTR) and the cystic fibrosis drug CPX using genome microarray analysis. *Mol. Med.* **5:** 753–767.
13. Virella-Lowell, I., J.D. Herlihy, B. Liu, *et al.* 2004. Effects of CFTR, interleukin-10, and Pseudomonas aeruginosa on gene expression profiles in a CF bronchial epithelial cell Line. *Mol. Ther.* **10:** 562–573.
14. Ollero, M., F. Brouillard & A. Edelman. 2006. Cystic fibrosis enters the proteomics scene: new answers to old questions. *Proteomics* **6:** 4084–4099.
15. Wetmore, D.R., E. Joseloff, J. Pilewski, *et al.* 2010. Metabolomic profiling reveals biochemical pathways and biomarkers associated with pathogenesis in cystic fibrosis cells. *J. Biol. Chem.* **285**: 30516–30522.
16. Henderson, M.J., O.V. Singh & P.L. Zeitlin. 2010. Applications of proteomic technologies for understanding the premature proteolysis of CFTR. *Expert Rev. Proteomics* **7:** 473–486.
17. Collawn, J.F., L. Fu & Z. Bebok. 2010. Targets for cystic fibrosis therapy: proteomic analysis and correction of mutant cystic fibrosis transmembrane conductance regulator. *Expert Rev. Proteomics* **7:** 495–506.
18. Gomes-Alves, P., M. Imrie, R.D. Gray, *et al.* 2010. SELDI-TOF biomarker signatures for cystic fibrosis, asthma and chronic obstructive pulmonary disease. *Clin. Biochem.* **43:** 168–177.
19. Pieroni, L., F. Finamore, M. Ronci, *et al.* 2011. Proteomics investigation of human platelets in healthy donors and cystic fibrosis patients by shotgun nUPLC-MSE and 2DE: a comparative study. *Mol. Biosyst.* **7:** 630–639.
20. Zabner, J., T.E. Scheetz, H.G. Almabrazi, *et al.* 2005. CFTR DeltaF508 mutation has minimal effect on the gene expression profile of differentiated human airway epithelia. *Am. J. Physiol. Lung Cell Mol. Physiol.* **289:** L545–L553.
21. Wright, J.M., C.A. Merlo, J.B. Reynolds, *et al.* 2006. Respiratory epithelial gene expression in patients with mild and severe cystic fibrosis lung disease. *Am. J. Respir. Cell Mol. Biol.* **35:** 327–336.
22. Wu, X., J.R. Peters-Hall, S. Ghimbovschi, *et al.* 2011. Glandular gene expression of sinus mucosa in chronic rhinosinusitis with and without cystic fibrosis. *Am. J. Respir. Cell Mol. Biol.* **45:** 525–533.
23. Srivastava, M., O. Eidelman, C. Jozwik, *et al.* 2006. Serum proteomic signature for cystic fibrosis using an antibody microarray platform. *Mol. Genet. Metab.* **87:** 303–310.
24. Charro, N., B.L. Hood, D. Faria, *et al.* 2011. Serum proteomics signature of cystic fibrosis patients: a complementary 2-DE and LC-MS/MS approach. *J. Proteomics* **74:** 110–126.
25. Jozwik, C.E., H.B. Pollard, M. Srivastava, *et al.* 2012. Antibody microarrays: analysis of cystic fibrosis. *Methods Mol. Biol.* **823:** 179–200.
26. Gharib, S.A., T. Vaisar, M.L. Aitken, *et al.* 2009. Mapping the lung proteome in cystic fibrosis. *J. Proteome Res.* **8:** 3020–3028.
27. Wolak, J.E., C.R. Esther, Jr. & T.M. O'Connell. 2009. Metabolomic analysis of bronchoalveolar lavage fluid from cystic fibrosis patients. *Biomarkers* **14:** 55–60.

28. Montuschi, P., D. Paris, D. Melck, *et al.* 2012. NMR spectroscopy metabolomic profiling of exhaled breath condensate in patients with stable and unstable cystic fibrosis. *Thorax* **67:** 222–228.
29. Robroeks, C.M., J.J. van Berkel, J.W. Dallinga, *et al.* 2010. Metabolomics of volatile organic compounds in cystic fibrosis patients and controls. *Pediatr. Res.* **68:** 75–80.
30. Collaco, J.M. & G.R. Cutting. 2008. Update on gene modifiers in cystic fibrosis. *Curr. Opin. Pulm. Med.* **14:** 559–566.
31. McDougal, K.E., D.M. Green, L.L. Vanscoy, *et al.* 2010. Use of a modeling framework to evaluate the effect of a modifier gene (MBL2) on variation in cystic fibrosis. *Eur. J. Hum. Genet.* **18:** 680–684.
32. Stanke, F., T. Becker, V. Kumar, *et al.* 2011. Genes that determine immunology and inflammation modify the basic defect of impaired ion conductance in cystic fibrosis epithelia. *J. Med. Genet.* **48:** 24–31.
33. Xu, Y., C. Liu, J.C. Clark & J.A. Whitsett. 2006. Functional genomic responses to cystic fibrosis transmembrane conductance regulator (CFTR) and CFTR(delta508) in the lung. *J. Biol. Chem.* **281:** 11279–11291.
34. Adib-Conquy, M., T. Pedron, A.F. Petit-Bertron, *et al.* 2008. Neutrophils in cystic fibrosis display a distinct gene expression pattern. *Mol. Med.* **14:** 36–44.
35. Painter, R.G., R.W. Bonvillain, V.G. Valentine, *et al.* 2008. The role of chloride anion and CFTR in killing of Pseudomonas aeruginosa by normal and CF neutrophils. *J. Leukoc. Biol.* **83:** 1345–1353.
36. Ratner, D. & C. Mueller. 2012. Immune responses in cystic fibrosis; are they intrinsically defective? *Am. J. Respir. Cell Mol. Biol.* Published ahead of print on March 8, 2012, doi: 10.1165/rcmb.2011-0399RT.
37. Wenk, M.R. 2010. Lipidomics: new tools and applications. *Cell* **143:** 888–895.
38. Quehenberger, O. & E.A. Dennis. 2011. The human plasma lipidome. *N. Engl. J. Med.* **365:** 1812–1823.
39. Bochkov, V.N., O.V. Oskolkova, K.G. Birukov, *et al.* 2010. Generation and biological activities of oxidized phospholipids. *Antioxid. Redox. Signal* **12:** 1009–1059.
40. Guerrera, I.C., G. Astarita, J.P. Jais, *et al.* 2009. A novel lipidomic strategy reveals plasma phospholipid signatures associated with respiratory disease severity in cystic fibrosis patients. *PLoS One* **4:** e7735.
41. Ollero, M., G. Astarita, I.C. Guerrera, *et al.* 2011. Plasma lipidomics reveals potential prognostic signatures within a cohort of cystic fibrosis patients. *J. Lipid. Res.* **52:** 1011–1022.
42. Sanak, M., A. Gielicz, K. Nagraba, *et al.* 2010. Targeted eicosanoids lipidomics of exhaled breath condensate in healthy subjects. *J. Chromatogr. B. Analyt. Technol. Biomed. Life Sci.* **878:** 1796–1800.
43. Serhan, C.N. 2010. Novel lipid mediators and resolution mechanisms in acute inflammation: to resolve or not? *Am. J. Pathol.* **177:** 1576–1591.
44. Massey, K.A. & A. Nicolaou. 2011. Lipidomics of polyunsaturated-fatty-acid-derived oxygenated metabolites. *Biochem. Soc. Trans.* **39:** 1240–1246.
45. Stables, M.J. & D.W. Gilroy. 2011. Old and new generation lipid mediators in acute inflammation and resolution. *Prog. Lipid Res.* **50:** 35–51.
46. Lundstrom, S.L., D. Balgoma, A.M. Wheelock, *et al.* 2011. Lipid mediator profiling in pulmonary disease. *Curr. Pharm. Biotechnol.* **12:** 1026–1052.
47. Ollero, M., I.C. Guerrera, G. Astarita, *et al.* 2011. New lipidomic approaches in cystic fibrosis. *Methods Mol. Biol.* **742:** 265–278.
48. Vaughan, W.J., F.T. Lindgren, J.B. Whalen & S. Abraham. 1978. Serum lipoprotein concentrations in cystic fibrosis. *Science* **199:** 783–786.
49. Freedman, S.D., P.G. Blanco, M.M. Zaman, *et al.* 2004. Association of cystic fibrosis with abnormalities in fatty acid metabolism. *N. Engl. J. Med.* **350:** 560–569.
50. Iuliano, L., R. Monticolo, G. Straface, *et al.* 2009. Association of cholesterol oxidation and abnormalities in fatty acid metabolism in cystic fibrosis. *Am. J. Clin. Nutr.* **90:** 477–484.
51. Carlstedt-Duke, J., M. Bronnegard, & B. Strandvik. 1986. Pathological regulation of arachidonic acid release in cystic fibrosis: the putative basic defect. *Proc. Natl. Acad. Sci. USA* **83:** 9202–9206.
52. Miele, L., E. Cordella-Miele, M. Xing, *et al.* 1997. Cystic fibrosis gene mutation (deltaF508) is associated with an intrinsic abnormality in Ca2+-induced arachidonic acid release by epithelial cells. *DNA Cell Biol.* **16:** 749–759.
53. Al-Turkmani, M.R., S.D. Freedman & M. Laposata. 2007. Fatty acid alterations and n-3 fatty acid supplementation in cystic fibrosis. *Prostaglandins Leukot. Essent. Fatty Acids* **77:** 309–318.
54. Rhodes, B., E.F. Nash, E. Tullis, *et al.* 2010. Prevalence of dyslipidemia in adults with cystic fibrosis. *J. Cyst. Fibros.* **9:** 24–28.
55. Strandvik, B. 2010. Fatty acid metabolism in cystic fibrosis. *Prostaglandins Leukot. Essent. Fatty Acids* **83:** 121–129.
56. Bravo, E., M. Napolitano, S.B. Valentini & S. Quattrucci. 2011. Neutrophil unsaturated fatty acid release by GM-CSF is impaired in cystic fibrosis. *Lipids Health Dis.* **9:** 129.
57. Njoroge, S.W., A.C. Seegmiller, W. Katrangi & M. Laposata. 2011. Increased Delta5- and Delta6-desaturase, cyclooxygenase-2, and lipoxygenase-5 expression and activity are associated with fatty acid and eicosanoid changes in cystic fibrosis. *Biochim. Biophys. Acta* **1811:** 431–440.
58. Njoroge, S.W., M. Laposata, W. Katrangi & A.C. Seegmiller. 2012. DHA and EPA reverse cystic fibrosis-related FA abnormalities by suppressing FA desaturase expression and activity. *J. Lipid Res.* **53:** 257–265.
59. Czerska, K., A. Sobczynska-Tomaszewska, D. Sands, *et al.* 2010. Prostaglandin-endoperoxide synthase genes COX1 and COX2—novel modifiers of disease severity in cystic fibrosis patients. *J. Appl. Genet.* **51:** 323–330.
60. Bonnans, C. & B.D. Levy. 2007. Lipid mediators as agonists for the resolution of acute lung inflammation and injury. *Am. J. Respir. Cell Mol. Biol.* **36:** 201–205.
61. Serhan, C.N., Y. Lu, S. Hong & R. Yang. 2007. Mediator lipidomics: search algorithms for eicosanoids, resolvins, and protectins. *Methods Enzymol.* **432:** 275–317.
62. Navarro-Xavier, R.A., J. Newson, V.L. Silveira, *et al.* 2010. A new strategy for the identification of novel molecules with

targeted proresolution of inflammation properties. *J. Immunol.* **184:** 1516–1525.

63. Visioli, F., E. Giordano, N.M. Nicod & A. Davalos. 2012. Molecular targets of omega 3 and conjugated linoleic Fatty acids—"micromanaging" cellular response. *Front. Physiol.* **3:** 1–11.
64. Higdon, A., A.R. Diers, J.Y. Oh, *et al.* 2012. Cell signalling by reactive lipid species: new concepts and molecular mechanisms. *Biochem. J.* **442:** 453–464.
65. Kiss, L., H. Schutte, W. Padberg, *et al.* 2010. Epoxyeicosatrienoates are the dominant eicosanoids in human lungs upon microbial challenge. *Eur. Respir. J.* **36:** 1088–1098.
66. Takai, D., T. Nagase & T. Shimizu. 2004. New therapeutic key for cystic fibrosis: a role for lipoxins. *Nat. Immunol.* **5:** 357–358.
67. Chiron, R., Y.Y. Grumbach, N.V. Quynh, *et al.* 2008. Lipoxin A(4) and interleukin-8 levels in cystic fibrosis sputum after antibiotherapy. *J. Cyst. Fibros.* **7:** 463–468.
68. Voynow, J.A. & B.K. Rubin. 2009. Mucins, mucus, and sputum. *Chest* **135:** 505–512.
69. Fahy, J.V. & B.F. Dickey. 2010. Airway mucus function and dysfunction. *N. Engl. J. Med.* **363:** 2233–2247.
70. Cox, M.J., M. Allgaier, B. Taylor, *et al.* 2010. Airway microbiota and pathogen abundance in age-stratified cystic fibrosis patients. *PLoS One* **5:** e11044.
71. Firoved, A.M. & V. Deretic. 2003. Microarray analysis of global gene expression in mucoid Pseudomonas aeruginosa. *J. Bacteriol.* **185:** 1071–1081.
72. Oberhardt, M.A., J.B. Goldberg, M. Hogardt & J.A. Papin. 2010. Metabolic network analysis of Pseudomonas aeruginosa during chronic cystic fibrosis lung infection. *J. Bacteriol.* **192:** 5534–5548.
73. Yang, J., K. Schmelzer, K. Georgi & B.D. Hammock. 2009. Quantitative profiling method for oxylipin metabolome by liquid chromatography electrospray ionization tandem mass spectrometry. *Anal. Chem.* **81:** 8085–8093.
74. Yang, J., J.P. Eisenrich, C.E. Cross, B.M. Morrissey & B.D. Hammock. 2012. Metabolomic profiling of regulatory lipid mediators in sputum from adult cystic fibrosis patients. *Free Rad. Biol. Med.* In press.
75. Moss, C.W., S.B. Samuels & R.E. Weaver. 1972. Cellular fatty acid composition of selected Pseudomonas species. *Appl. Microbiol.* **24:** 596–598.
76. Sato, H., J.B. Feix, C.J. Hillard & D.W. Frank. 2005. Characterization of phospholipase activity of the Pseudomonas aeruginosa type III cytotoxin, ExoU. *J. Bacteriol.* **187:** 1192–1195.
77. Vance, R.E., S. Hong, K. Gronert, *et al.* 2004. The opportunistic pathogen Pseudomonas aeruginosa carries a secretable arachidonate 15-lipoxygenase. *Proc. Natl. Acad. Sci. USA* **101:** 2135–2139.
78. Martinez, E., M. Hamberg, M. Busquets, *et al.* 2010. Biochemical characterization of the oxygenation of unsaturated fatty acids by the dioxygenase and hydroperoxide isomerase of Pseudomonas aeruginosa 42A2. *J. Biol. Chem.* **285:** 9339–9345.
79. Bahl, C.D., C. Morisseau, J.M. Bomberger, *et al.* 2010. Crystal structure of the cystic fibrosis transmembrane conductance regulator inhibitory factor Cif reveals novel active-site features of an epoxide hydrolase virulence factor. *J. Bacteriol.* **192:** 1785–1795.
80. Mayer-Hamblett, N., M.L. Aitken, F.J. Accurso, *et al.* 2007. Association between pulmonary function and sputum biomarkers in cystic fibrosis. *Am. J. Respir. Crit. Care Med.* **175:** 822–828.
81. Konstan, M.W., J.S. Wagener, A. Yegin, *et al.* 2010. Design and powering of cystic fibrosis clinical trials using rate of FEV(1) decline as an efficacy endpoint. *J. Cyst. Fibros.* **9:** 332–338.
82. Liou, T.G., E.P. Elkin, D.J. Pasta, *et al.* 2010. Year-to-year changes in lung function in individuals with cystic fibrosis. *J. Cyst. Fibros.* **9:** 250–256.
83. Jiang, Q., X. Yin, M.A. Lill, *et al.* 2008. Long-chain carboxychromanols, metabolites of vitamin E, are potent inhibitors of cyclooxygenases. *Proc. Natl. Acad. Sci. USA* **105:** 20464–20469.
84. Keen, C., A.C. Olin, S. Eriksson, *et al.* 2010. Supplementation with fatty acids influences the airway nitric oxide and inflammatory markers in patients with cystic fibrosis. *J. Pediatr. Gastroenterol. Nutr.* **50:** 537–544.
85. Rice, T.W., A.P. Wheeler, B.T. Thompson, *et al.* 2011. Enteral omega-3 fatty acid, gamma-linolenic acid, and antioxidant supplementation in acute lung injury. *JAMA* **306:** 1574–1581.
86. Sabater, J., J.R. Masclans, J. Sacanell, *et al.* 2011. Effects of an omega-3 fatty acid-enriched lipid emulsion on eicosanoid synthesis in acute respiratory distress syndrome (ARDS): A prospective, randomized, double-blind, parallel group study. *Nutr. Metab. (Lond)* **8:** 22.
87. van der Meij, B.S., M.A. van Bokhorst-de van der Schueren, J.A. Langius, *et al.* 2011. n-3 PUFAs in cancer, surgery, and critical care: a systematic review on clinical effects, incorporation, and washout of oral or enteral compared with parenteral supplementation. *Am. J. Clin. Nutr.* **94:** 1248–1265.
88. Guilbault, C., G. Wojewodka, Z. Saeed, *et al.* 2009. Cystic fibrosis fatty acid imbalance is linked to ceramide deficiency and corrected by fenretinide. *Am. J. Respir Cell Mol. Biol.* **41:** 100–106.
89. Rossi, A., C. Pergola, A. Koeberle, *et al.* 2010. The 5-lipoxygenase inhibitor, zileuton, suppresses prostaglandin biosynthesis by inhibition of arachidonic acid release in macrophages. *Br. J. Pharmacol.* **161:** 555–570.
90. Liu, J.Y., J. Yang, B. Inceoglu, *et al.* 2010. Inhibition of soluble epoxide hydrolase enhances the anti-inflammatory effects of aspirin and 5-lipoxygenase activation protein inhibitor in a murine model. *Biochem. Pharmacol.* **79:** 880–887.
91. Ribeiro, C.M., H. Hurd, Y. Wu, *et al.* 2009. Azithromycin treatment alters gene expression in inflammatory, lipid metabolism, and cell cycle pathways in well-differentiated human airway epithelia. *PLoS One* **4:** e5806.
92. Planaguma, A., M.A. Pfeffer, G. Rubin, *et al.* 2010. Lovastatin decreases acute mucosal inflammation via 15-epi-lipoxin A4. *Mucosal. Immunol.* **3:** 270–279.
93. Kaddurah-Daouk, R., R.A. Baillie, H. Zhu, *et al.* 2010. Lipidomic analysis of variation in response to simvastatin in the Cholesterol and Pharmacogenetics Study. *Metabolomics* **6:** 191–201.

94. Brennan, S. 2007. Revisiting alpha1-antitrypsin therapy in cystic fibrosis: can it still offer promise? *Eur. Respir. J.* **29:** 229–230.

95. Spite, M., L.V. Norling, L. Summers, *et al.* 2009. Resolvin D2 is a potent regulator of leukocytes and controls microbial sepsis. *Nature* **461:** 1287–1291.

Ann. N.Y. Acad. Sci. ISSN 0077-8923

ANNALS OF THE NEW YORK ACADEMY OF SCIENCES
Issue: *Environmental Stressors in Biology and Medicine*

Nitric oxide signaling in the brain: translation of dynamics into respiration control and neurovascular coupling

João Laranjinha, Ricardo M. Santos, Cátia F. Lourenço, Ana Ledo, and Rui M. Barbosa
Center for Neurosciences and Cell Biology, and Faculty of Pharmacy, University of Coimbra, Coimbra, Portugal

Address for correspondence: João Laranjinha, Faculty of Pharmacy and Center for Neurosciences and Cell Biology, University of Coimbra, Health Sciences Campus, Azinhaga de Santa Comba, 3000-548 Coimbra, Portugal. laranjin@ci.uc.pt

The understanding of the unorthodox actions of neuronal-derived nitric oxide (•NO) in the brain has been constrained by uncertainties regarding its quantitative profile of change in time and space. As a diffusible intercellular messenger, conveying information associated with its concentration dynamics, both the synthesis via glutamate stimulus and inactivation pathways determine the profile of •NO concentration change. *In vivo* studies, encompassing the real-time measurement of •NO concentration dynamics have allowed us to gain quantitative insights into the mechanisms inherent to •NO-mediated signaling pathways. It has been of particular interest to study the diffusion properties and half-life, the interplay between •NO and O_2 and the ensuing functional consequences for regulation of O_2 consumption, the role of vasculature in shaping •NO signals *in vivo,* and the mechanisms that are responsible for •NO to achieve the coupling between glutamatergic neuronal activation and local microcirculation.

Keywords: nitric oxide diffusion; brain; oxygen; neurovascular coupling; *in vivo*

Introduction

Nitric oxide (•NO) is a messenger that mediates a variety of physiologic processes in the brain, such as neurotransmitter release, neuronal excitability, long-term potentiation, and neurovascular coupling. However, this free radical has also been implicated in excitotoxic pathways underlying neurological disorders such as Alzheimer's and Parkinson's diseases.[1,2] Unusual physical and chemical properties underlie the diversity of •NO actions. Remarkably, the ability of •NO to diffuse across membranes because of its small size and hydrophobicity, together with its radical nature, greatly accounts for the wide range of interactions of this molecule with biological targets in a concentration and redox environment–dependent fashion.[3]

Given the concentration dynamics dependence of •NO actions, it is critical to monitor with high temporal and spatial resolution •NO concentration profiles in the brain in connection with functional actions.

During the last few years, we have studied •NO activity in the brain, performing rapid (subsecond) and sensitive measurements of •NO in intact brain tissue using •NO-selective microelectrodes. This approach has allowed us to gain insights into the mechanisms that determine •NO bioactivity in the brain at two different levels: regulation of •NO concentration dynamics, namely by •NO diffusion and inactivation, and the functional outcome of endogenous •NO profiles on the regulation of O_2 concentration and local blood flow (Fig. 1).

Nitric oxide diffusion and half-life

Together with the temporal and spatial properties of •NO production in a given brain region, two kinetic parameters dictate the shape of •NO signals in tissue: the rate of •NO diffusion and its half-life.

It has been proposed that •NO is able to freely diffuse in tissues because of its small size and hydrophobic nature.[4] Accordingly, values of •NO diffusion coefficient (D_{NO}) determined in aqueous solutions that range from 2 to 4.5×10^{-5} cm^2/s [5–7] have been used to model •NO concentration dynamics in the brain.[8–10] However, quantitative data

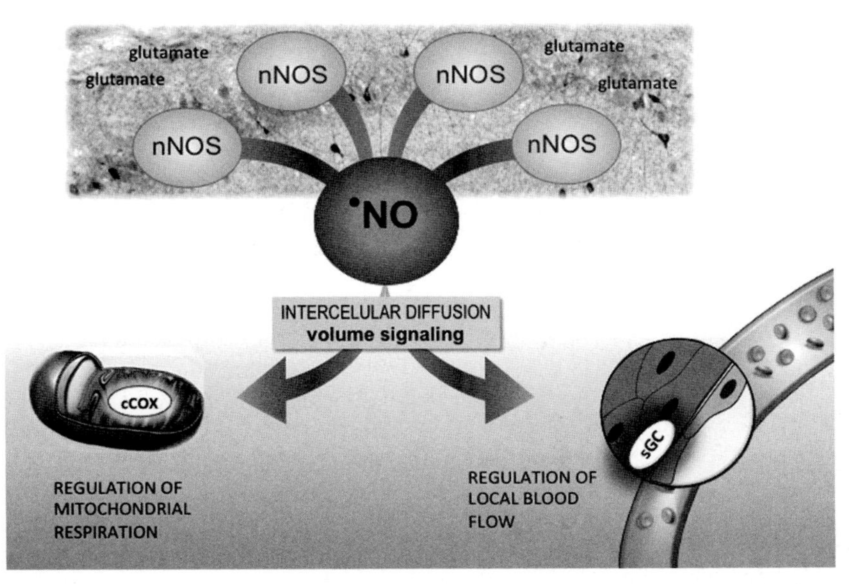

Figure 1. Modulation of oxygen tension in the brain by nitric oxide. General scheme depicting the functional impact of neuronal-derived •NO concentration dynamics into two critical processes: cellular respiration, via competition with oxygen for cytochrome c oxidase; and regulation of local blood flow (neurovascular coupling), via interaction with soluble guanylate cyclase in smooth muscle cells of the arteriole wall.

regarding •NO diffusion in the brain *in vivo* have been very limited due to difficulties in direct •NO measurement in intact tissue with high sensitivity and spatiotemporal resolution. These technical difficulties have also undermined a detailed study of the kinetics of •NO inactivation *in vivo*. Studies performed in isolated cell preparations of brain and liver extrapolated values for *in vivo* of around 0.1 s, assuming apparent first-order kinetics of NO decay,[11,12] in accordance with the half-life obtained in heart muscle.[13] In intact brain tissue, indirect measurements based on cGMP accumulation across cerebellar brain slices found Michaelis Menten kinetics of NO inactivation imposing a •NO half-life of 0.01 s for •NO concentrations bellow 10 nM.[14] However, a longer half-life of 0.6 s was estimated in organotypic slices.[15] These data demonstrate high uncertainty in the estimation of NO half-life and, importantly, none of the above studies were performed in the brain *in vivo*.

We have recently obtained novel quantitative data on •NO diffusion and half-life in the rat brain *in vivo*.[16] The experimental approach was based on the use of •NO-selective carbon fiber microelectrodes to monitor •NO changes upon local application of small volumes (low nanoliter range) of the •NO solution from a micropipette closely located at 200–350 μm from the microelectrode. These arrays were stereotaxically inserted in the brain of anesthetized rats.

The resulting electrochemical signals (shown in Fig. 2) were fitted to a diffusion/inactivation equation, which allowed the estimation of the •NO diffusion coefficient and half-life (Fig. 2). We found that •NO is highly diffusible and short lived in the brain, having an effective diffusion coefficient (D_{NO}*) of 2.2×10^{-5} cm^2/s and a half-life of 0.64 s in the rat brain cortex. We further tested the concepts of free, hindered, and enhanced diffusion to better characterize •NO diffusion in the brain. The D_{NO} obtained following local application of •NO in the agarose gel, used to evaluate •NO free diffusion, was 2.6×10^{-5} cm^2/s, only 14% higher than the *in vivo* D_{NO}*, suggesting that •NO can freely diffuse in the brain. However, subsequent experiments indicated that •NO diffusion in brain tissue is considerably faster than that of a molecule of similar size that remains in the extracellular space (nitrite), supporting the fact that •NO is able to diffuse across cell membranes, and thereby its diffusion is not retarded by the tortuosity of the extracellular media. This idea indicates that a large fraction of the •NO diffusion pathway in the brain is intracellular.

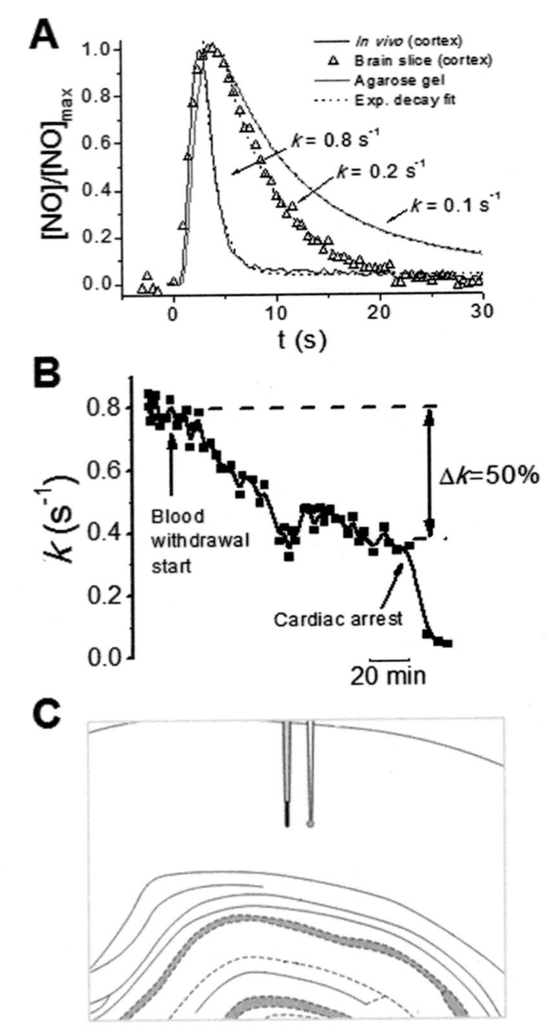

Figure 2. Nitric oxide inactivation in the brain. (A) Representative normalized •NO signals obtained upon local application of small volumes (10 nL) of •NO solution in the brain *in vivo*, in brain slices and in agarose (used as a control for the contribution of NO diffusion to signal decay). First-order decay rate constants (k) of each signal decay are indicated. (B) Graph showing a representative decrease in k following hemorrhagic shock, induced by progressive withdrawal of 6 mL of blood over 40 min (adapted from Ref. 49). (C) Schematic representation of recording site *in vivo*.

Nevertheless, we also observed that •NO diffusion is hindered by macromolecular crowding (evaluated in BSA-supplemented gel experiments), indicating that a homogeneous •NO diffusion in the tissue, encompassing the crowded intracellular microenvironment, does not provide a reasonable conceptual explanation for the similarity between the D_{NO}* obtained *in vivo* and the D_{NO} found in the agarose gel. Thus, as previously suggested for O_2,[17,18] it seems plausible that physiological processes facilitate •NO diffusion in nervous tissue. A likely candidate is a •NO partition in hydrophobic media, such as cell membranes and myelin sheaths, which might constitute low-resistance pathways that facilitate •NO diffusion in nervous tissue and consequently increase the •NO diffusion radius.

The •NO diffusion radius is an important piece of information with great relevance regarding •NO bioactivity in the brain, allowing a better understanding of how this small molecule may act as a volume-signaling messenger. We directly addressed this topic in acute hippocampal slices, by performing measurements of endogenously produced •NO via the activation of NMDA-type glutamate receptors, the main pathway for •NO production in the brain.[48] We found that, for the activation of multiple NOSs contained within a 50 μm radius, the maximal diffusional spread of •NO in the CA1 subregion was 400 μm.[19] These results and those of other authors[4,20–22] demonstrated that •NO produced by multiple sources contained within a finite volume can diffuse a large distance beyond that same volume and integrate the activity of multiple cells and cell types (neurons, asctrocytes, endothelial cells, etc.) without the requirement of physical coupling or even proximity to the •NO-producing cell. The high D_{NO}* that we found in the brain *in vivo* agrees with this notion. However, given that both •NO diffusion and •NO half-life determine •NO diffusion radius, the results in brain slices may be overestimating the •NO diffusion radius compared with *in vivo* conditions. As discussed in the forthcoming sections, we have found that •NO half-life is considerably longer (about threefold higher) in brain slices than *in vivo*, thus suggesting a smaller •NO diffusion radius *in vivo*. Regardless of the actual •NO diffusion radius *in vivo*, these results clearly portray •NO as a unique and unorthodox signaling molecule in the brain.

The nitric oxide and oxygen interplay with consequences for cellular metabolism

•NO and O_2 are hydrophobic and diffusible gases with high biological relevance. The intricate interplay between them leads to a reciprocal regulation of usage, bioavailability, and bioactivity in the brain. These molecules can react between themselves (auto-oxidation), producing nitrite in aqueous solution. At low concentrations of •NO found *in vivo*, this reaction is too slow to account for significant •NO consumption, but the high partition of these molecules in the hydrophobic phase of biomembranes accelerates the reaction, partially accounting for •NO consumption in tissues.[23] In fact, NO consumption by autooxidation in mitochondrial membranes was proposed to modulate NO inhibition of cellular respiration.[24]

Yet, the interplay between •NO and O_2 is not limited to autooxidation. The better characterized biochemical target for •NO is soluble guanylate cyclase,[25] but cytochrome c oxidase (CcO) has emerged as a critical mediator of •NO biological activity. Although CcO was known to embark in redox interactions with •NO,[26] the inhibition of CcO by •NO was only demonstrated in the mid-1990s in isolated mitochondria[27,28] and synaptosomes.[29] Recently, we were able to contribute to the better understanding of the •NO/O_2 interplay in the brain by evaluating the impact of endogenous •NO (resulting from the activation of NMDA-type glutamate receptors) on O_2 consumption in hippocampal slices.[30] Our results showed that, in a complex biological preparation where the cytoarchitectural integrity and neuronal circuitry are maintained, submicromolar concentrations of •NO inhibit tissue O_2 consumption in the CA1 subregion. These findings point toward a direct relationship between •NO concentration and extent of inhibition and substantiate the hypothesis of a sparing capacity of CcO[31] based on the observation that a threshold concentration of •NO is needed to produce a net effect on tissue O_2 consumption.

CcO is the terminal complex in the mitochondrial respiratory chain, receiving electrons needed for O_2 reduction from cytochrome c. •NO competes with O_2 for binding to the fully reduced heme-copper active site,[27,29] resulting in an increase in the K_m for O_2. This reversible inhibition has been shown to occur at low nM concentrations of •NO such as those observed upon stimulation of NMDA type of glutamate receptors in the hippocampus.[19,30] Remarkably, the potency of •NO significantly increases when O_2 concentration is decreased or when the enzyme turnover increases.[32] These observations imply that under conditions of stimulated neuronal activity which demand increased O_2 consumption, or under hypoxic conditions, the •NO inhibition of CcOx becomes enhanced. Accordingly, recent evidence indicates that inhibition of CcOx might play a role in the adjustment of tissue oxygen consumption at low $[O_2]$.[33] •NO can also bind and react with the oxidized CcO at the Cu_B^{2+} in an uncompetitive inhibition mechanism, which is reversed resulting in •NO oxidation to NO_2^- and CcO reduction.[34] Consequently, the binding of •NO to CcO may serve two purposes: regulation of cellular respiration rate and the shaping of •NO concentration dynamics by •NO inactivation.

The inhibition of CcO by •NO is also relevant in the regulation of physiological processes including activation of the hypoxia-inducible factor,[35] neurovascular coupling,[36] increased diffusion of O_2 and thus increased oxygenation to cells more distant from capillaries.[36,37] Reduced electron flow at CcO due to •NO inhibition can increase $O_2^{-\bullet}$ production due to electron leakage at complexes I and III.[38–41] In this way $O_2^{-}\bullet$ may contribute to •NO inactivation via radical–radical interaction at the diffusion limit (1.9×10^{10}/M/s),[42] yielding peroxynitrite ($ONOO^-$), a potent oxidant and nitrating species. Increased protein nitration, including the mitochondrial respiratory complexes,[43] has been described in aged rats and in models of Alzheimer's disease and in the brains of Alzheimer's patients.[44,45] Thus, the interplay between •NO and O_2 in brain tissue must be a fine-tuned process.

A group of proteins from the globin family expressed in the brain have also been shown to react with •NO when found in the O_2-bound ferrous-heme state ($k \approx 10^7$/M/s). Although the physiological role of proteins such as neuroglobin, cytoglobin, and hemoglobin[46–48] is yet unclear in nervous tissue, they may act as O_2-dependent •NO scavengers.

In accordance with a role of O_2-dependent •NO inactivation in the regulation of •NO dynamics in the brain, we found that the decay of electrochemical signals, recorded with •NO-selective microelectrodes upon local application of •NO in the brain of anesthetized rats, was slightly dependent on tissue oxygen tension. The decay rate constant of •NO signals decreased 20% under anoxia and increased to a similar extent during hyperoxia[49] supporting a small but significant contribution of O_2-dependent processes for •NO inactivation in the brain *in vivo*, though the exact mechanisms remain to be explored.

Nitric oxide signaling along the trisynaptic loop in the hippocampus and the role of vasculature in shaping the volume signaling

The hippocampus is a heterogenous structure, both cytoarchitecturally and functionally, exhibiting regional specificity for encoding certain types of information, such as novelty detection, separation patterns, and retrieval.[50]

•NO has a critical role in several models of synaptic plasticity, including long-term potentiation (LTP) in the hippocampus, and may also be implicated in glutamate NMDA receptor-dependent LTP in all subregions, although the phenomenon appears to occur differentially throughout the trisynaptic loop. This observation raises the question of whether •NO concentration dynamics are heterogeneous along the trisynaptic loop. To help answer this question, it is important to consider that •NO may act either in a synapse-specific manner or as a volume-signaling messenger. Theoretical studies have shown that when produced in an isolated synapse, the •NO local concentration decays steeply a few micrometers away from its site of synthesis,[9,51] therefore, likely acting in a synapse-specific manner. However, because •NO rapidly diffuses in brain tissue with little constraint,[16] individual sources of •NO that become activated simultaneously might summate[9] originating volume signals that are distinct in amplitude, spatial characteristics, and time course from those resulting from isolated synaptic activity. This type of intercellular signaling is remarkably different from conventional neurotransmission and likely encompasses biological effects distinct from those encoded by •NO signals from single synapses. For instance, whereas the former seems suitable to modulate pre- and postsynaptic responses in particular synapses,[51] a volume signal might additionally induce more global responses such as the modulation of O_2 availability in the tissue either through an increase in local blood flow or partial inhibition of cellular respiration. It is worth noting that given the brain's strong dependence on O_2, the regulation of these intermingled phenomena, cellular respiration, and local blood flow is critical for the structural and functional integrity of the brain. Therefore, we have studied the concentration dynamics of NO in connection with those phenomena, as summarized here.

Although the relative contribution of synapse-specific versus •NO volume signaling for processes like LTP remains unclear, we observed that, indeed, •NO volume signals are heterogeneous within the hippocampus. Endogenous volume signals were induced by local stimulation of a volume of tissue with glutamate ejected from a micropipette located close to a •NO-selective microelectrode. We observed that upon glutamatergic activation, the resulting •NO volume signals *in vivo* possess a longer time course in the *CA1* subregion compared with the dentate gyrus[49] (Fig. 3) and also within the *CA3* subregion (unpublished data). Although the peak concentration was not dramatically different, signals last longer in the *CA1* subregion than in the dentate gyrus (86 ± 3 s vs. 58 ± 9 s), thus resulting in a higher output of •NO, and likely a larger •NO volume signaling, in the former subregion of the hippocampus. The molecular mechanisms underlying the heterogeneous •NO concentrations along the hippocampal trisynaptic loop remain to be explored. It can be anticipated that differences in the expression and/or regulation of neuronal isoform of nitric oxide synthase (nNOS) or ionotropic glutamate receptors may critically contribute to the shape profile of •NO production.

Additionally, the downstream mechanisms that govern •NO inactivation and diffusion may also contribute to modulate the •NO concentration dynamics. The kinetics of •NO clearance from the vicinity of an individual source is mainly determined by rapid •NO diffusion to the neighboring tissue, which overcomes the contribution of •NO inactivation (unless an extremely short •NO half-life in the millisecond range is assumed).[9,14] However, when multiple •NO sources summate, •NO inactivation (which determines •NO half-life) plays an important role in shaping the spatial, temporal, and amplitude properties of the resulting volume signal. Interestingly, mathematical simulations have shown that cooperativity between individual •NO sources is significantly decreased for •NO half-lives below 500 milliseconds.[9] According to these theoretical studies, the •NO half-life of 0.64 s that we have determined in the rat brain cortex *in vivo* appears to be fine-tuned to ensure strong cooperativity between •NO sources in addition to maintaining physiologic •NO concentrations in tissue.[16]

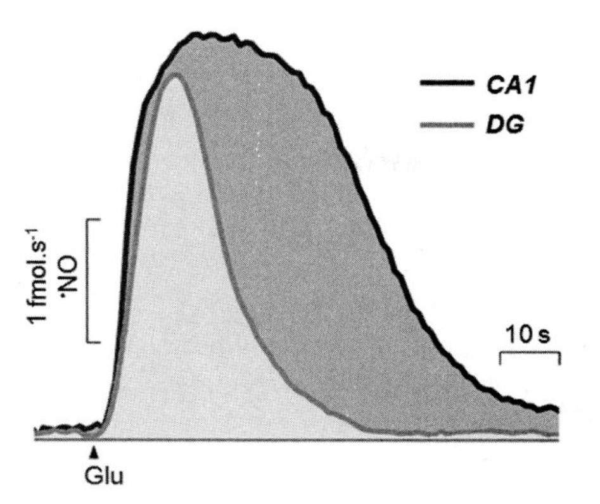

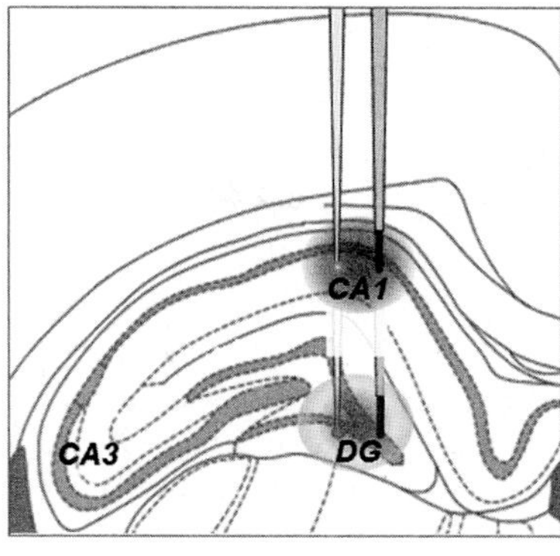

Figure 3. *In vivo* concentration dynamics of nitric oxide. Left panel: concentration dynamics of NO production upon glutamatergic activation in the CA1 subregion and dentate gyrus (DG) of the rat hippocampus. Glutamate (20 mM) was locally applied by pressure ejection at the time indicated by the arrow. Right panel: schematic representation of recording sites.

The effect of •NO half-life on the spatiotemporal properties of •NO signals in tissue, and consequently •NO signaling mechanisms and effects, underlies the importance of identifying the mechanisms of •NO inactivation in the brain. To study the mechanisms that govern •NO inactivation, we have recorded •NO signals using •NO-selective microelectrodes following local application of small volumes of exogenous •NO in the brain.[49] The decay of •NO signals obtained by this approach was very sensitive to vascular changes *in vivo*, namely global ischemia and hemorrhagic shock (Fig. 2B shows a 50% decrease in first-order k values of signal decay during hemorrhagic shock) and more than threefold faster than in brain slices, which lack functional vasculature (the temporal profile and first-order k values of signal decay in different media are shown in Fig. 2). Comparatively, modulation of O_2 tension in the brain *in vivo* caused only small changes in •NO decay, compared with impairment in vascular function, thus demonstrating that •NO scavenging by circulating red blood cells constitutes the major •NO inactivation pathway in the brain. Our data also suggest an inverse relationship between vascular density and •NO diffusion radius.[49]

Thus, the vascular network seems critical for the regulation of •NO-mediated volume signaling in the brain. Accordingly, vascular density is reported to be heterogeneous within the hippocampus, with total vessel area being higher in *dentate gyrus* than in *CA1* and *CA3* subregions,[52] which might partially account for the differences that we observed in •NO volume signals between these regions.

Regardless of the mechanisms underlying the heterogeneous profiles of •NO concentration within the hippocampus, the differences observed in the final output of •NO upon glutamatergic activation can have important implications. It may, for instance, reverberate in the selective vulnerability of the *CA1* subregion to pathological conditions, such as excitotoxicity and ischemic injury.[53]

Functional implications of •NO diffusion for neurovascular coupling

Soon after the identification of •NO as an endothelial-derived relaxing factor and as a neuromodulator, it emerged as a liable candidate to mediate the neurovascular coupling,[54] an active mechanism enlarging vessel diameter in response to the rising of metabolic demands imposed by neuronal activity that is of vital importance for the brain to preserve its structural and functional integrity.[55] •NO, due to its peculiar properties, is well suited to mediate the coupling between neuronal activity and cerebral blood flow given that it is a potent vasodilator, it is released during enhanced neuronal activity resulting from glutamatergic activation, and, as previously discussed, it is also highly diffusible. The suitability of •NO to mediate the neurovascular coupling is further supported by the integration of two observations, as previously mentioned, that the NO diffusional field in hippocampal slices is approximately 400 μm,[19] and the mean distance between NADPH-diaphorase positive fibers in the CA1 pyramidal layer and blood vessels, particularly arterioles, is around 150 μm[56]—thus lower than the •NO diffusional spread in the slices.

Following the formulation of the hypothesis that •NO might bridge neural activity with changes in local microcirculation, some evidence has been collected supporting its involvement the neurovascular coupling process, mainly based on observations that vasoactive actions of glutamate were reduced by NOS inhibitors. Indeed, the vascular responses elicited by NMDA and glutamate were restrained by NOS inhibitors both in hippocampal brain slices[56] and *in vivo* using a cranial window approach.[57–59] Similar results were reported when somatosensory stimulation[60] or afferent electric stimulation[61,62] was used to evoke neurovascular coupling in cerebral and cerebellar cortex, respectively.

In addition to reports supporting •NO involvement in the neurovascular coupling in regions such as the cerebellum, cerebral cortex, and hippocampus, other reports suggest the opposite or delegate •NO to play a secondary role as a modulator of other pathways, thereby mediating the process. Studies using somatosensory stimulation reported the ineffectual action of NOS inhibitors over CBF increases, suggesting the lack of •NO-mediated effects.[63–65] Other studies have reported that although •NO is required for neurovascular coupling, at least in somatosensory cortex, it does not directly mediate the neuron to vessels signals,[66] but rather acts as a modulator of other pathways associated to the vasoactive arachidonic acid metabolites.[67]

Thus, despite the likelihood that neuronal-derived •NO is critically involved in the regulation of cerebral blood flow, direct and real-time evidence is still missing.[68] The concept of proof requires the establishment of a coupling between neuronal-derived •NO and changes in local blood flow for space, time, and amplitude. Therefore, a quantitative, real-time, simultaneous, and *in vivo* measurement of •NO concentration dynamics and changes in blood flow is required. The possibility of measuring directly •NO in a real-time fashion *in vivo* gives us an important advantage in providing clues about the involvement of •NO in neurovascular coupling, particularly in the hippocampus, a brain region that has been comparatively ignored in regarding this issue. Through the development of an array consisting of the association of •NO microelectrodes and laser Doppler flowmetry with a predefined geometry, which allows probing simultaneously the dynamics of •NO and cerebral local flow changes, we expect to be able to establish the temporal, amplitude, and spatial association of both events. We are collecting robust data supporting a critical involvement of neuron-derived •NO in the neurovascular coupling in the hippocampus (in publication).

It is envisaged, however, that the dual interaction that likely occurs between events supports the hypothesis of a highly and intrinsically controlled mechanism for mediating neurovascular coupling. While •NO promotes an increase in cerebral blood flow, in turn, the increase in the cerebral blood flow, by inactivating •NO, may contribute to shaping the •NO signal.

Thus, this work may establish the basis of a novel mechanism by which local cerebral blood flow responds to neuronal stimulation—a connection mediated by diffusion via neuronal-derived •NO—regardless of the intermediacy of cellular players, such as astrocytes.

Acknowledgments

This work was supported by Grants PTDC/SAU-NEU/108992/2008 and PTDC/SAU-NEU/103538/2008, and PTDC/SAU-BEB/ 103228/2008 from FCT (Portugal).

Conflicts of interest

The authors declare no conflicts of interest.

References

1. Attwell, D. *et al.* 2010. Glial and neuronal control of brain blood flow. *Nature* **468:** 232–243.
2. Steinert, J.R., T. Chernova & I. D. Forsythe. 2010. Nitric oxide signaling in brain function, dysfunction, and dementia. *Neuroscientist* **16:** 435–452.
3. Hill, B.G. *et al.* 2010. What part of NO don't you understand? Some answers to the cardinal questions in nitric oxide biology. *J. Biol. Chem.* **285:** 19699–19704.
4. Lancaster, J.R., Jr. 1997. A tutorial on the diffusibility and reactivity of free nitric oxide. *Nitric Oxide* **1:** 18–30.
5. Denicola, A. *et al.* 1996. Nitric oxide diffusion in membranes determined by fluorescence quenching. *Arch. Biochem. Biophys.* **328:** 208–212.
6. Malinski, T. *et al.* 1993. Diffusion of nitric oxide in the aorta wall monitored in situ by porphyrinic microsensors. *Biochem. Biophys. Res. Commun.* **193:** 1076–1082.
7. Zacharia, I.G. & W.M. Deen. 2005. Diffusivity and solubility of nitric oxide in water and saline. *Ann. Biomed. Eng.* **33:** 214–222.

8. Philippides, A., P. Husbands & M. O'Shea. 2000. Four-dimensional neuronal signaling by nitric oxide: a computational analysis. *J. Neurosci.* **20:** 1199–1207.
9. Philippides, A. *et al.* 2005. Modeling cooperative volume signaling in a plexus of nitric-oxide-synthase-expressing neurons. *J. Neurosci.* **25:** 6520–6532.
10. Steinert, J.R. *et al.* 2008. Nitric oxide is a volume transmitter regulating postsynaptic excitability at a glutamatergic synapse. *Neuron* **60:** 642–656.
11. Griffiths, C. & J. Garthwaite. 2001. The shaping of nitric oxide signals by a cellular sink. *J. Physiol.* **536:** 855–862.
12. Thomas, D.D. *et al.* 2001. The biological lifetime of nitric oxide: implications for the perivascular dynamics of NO and O2. *Proc. Natl. Acad. Sci. USA* **98:** 355–360.
13. Kelm, M. & J. Schrader. 1990. Control of coronary vascular tone by nitric oxide. *Circ. Res.* **66:** 1561–1575.
14. Hall, C.N. & J. Garthwaite. 2006. Inactivation of nitric oxide by rat cerebellar slices. *J. Physiol.* **577:** 549–567.
15. Hall, C.N. & D. Attwell. 2008. Assessing the physiological concentration and targets of nitric oxide in brain tissue. *J. Physiol.* **586:** 3597–3615.
16. Santos, R.M. *et al.* 2011. Evidence for a pathway that facilitates nitric oxide diffusion in the brain. *Neurochem. Int.* **59:** 90–96.
17. Bentley, T.B. & R.N. Pittman. 1997. Influence of temperature on oxygen diffusion in hamster retractor muscle. *Am. J. Physiol.* **272:** H1106–H1112.
18. Sidell, B.D. 1998. Intracellular oxygen diffusion: the roles of myoglobin and lipid at cold body temperature. *J. Exp. Biol.* **201:** 1119–1128.
19. Ledo, A. *et al.* 2005. Concentration dynamics of nitric oxide in rat hippocampal subregions evoked by stimulation of the NMDA glutamate receptor. *Proc. Natl. Acad. Sci. USA* **102:** 17483–17488.
20. Lancaster, J.R. Jr. 1994. Simulation of the diffusion and reaction of endogenously produced nitric oxide. *Proc. Natl. Acad. Sci. USA* **91:** 8137–8141.
21. Wood, J. & J. Garthwaite. 1994. Models of the diffusional spread of nitric oxide: implications for neural nitric oxide signalling and its pharmacological properties. *Neuropharmacology* **33:** 1235–1244.
22. Lancaster, J.R., Jr. 1996. Diffusion of free nitric oxide. *Methods Enzymol.* **268:** 31–50.
23. Liu, X. *et al.* 1998. Accelerated reaction of nitric oxide with O2 within the hydrophobic interior of biological membranes. *Proc. Natl. Acad. Sci. USA* **95:** 2175–2179.
24. Shiva, S. *et al.* 2001. Nitric oxide partitioning into mitochondrial membranes and the control of respiration at cytochrome c oxidase. *Proc. Natl. Acad. Sci. USA* **98:** 7212–7217.
25. Poulos, T.L. 2006. Soluble guanylate cyclase. *Curr. Opin. Struct. Biol.* **16:** 736–743.
26. Brudvig, G.W., T.H. Stevens & S.I. Chan. 1980. Reactions of nitric oxide with cytochrome c oxidase. *Biochemistry* **19:** 5275–5285.
27. Cleeter, M.W. *et al.* 1994. Reversible inhibition of cytochrome c oxidase, the terminal enzyme of the mitochondrial respiratory chain, by nitric oxide. Implications for neurodegenerative diseases. *FEBS Lett.* **345:** 50–54.
28. Schweizer, M. & C. Richter. 1994. Nitric oxide potently and reversibly deenergizes mitochondria at low oxygen tension. *Biochem. Biophys. Res. Commun.* **204:** 169–175.
29. Brown, G.C. & C.E. Cooper. 1994. Nanomolar concentrations of nitric oxide reversibly inhibit synaptosomal respiration by competing with oxygen at cytochrome oxidase. *FEBS Lett.* **356:** 295–298.
30. Ledo, A. *et al.* 2010. Dynamic and interacting profiles of *NO and O2 in rat hippocampal slices. *Free Radic. Biol. Med.* **48:** 1044–1050.
31. Dranka, B.P., B.G. Hill & V.M. Darley-Usmar. 2010. Mitochondrial reserve capacity in endothelial cells: The impact of nitric oxide and reactive oxygen species. *Free Radic. Biol. Med.* **48:** 905–914.
32. Cooper, C.E. & C. Giulivi. 2007. Nitric oxide regulation of mitochondrial oxygen consumption II: Molecular mechanism and tissue physiology. *Am. J. Physiol. Cell Physiol.* **292:** C1993–C2003.
33. Victor, V.M. *et al.* 2009. Regulation of oxygen distribution in tissues by endothelial nitric oxide. *Circ. Res.* **104:** 1178–1183.
34. Cooper, C.E. *et al.* 1997. Nitric oxide ejects electrons from the binuclear centre of cytochrome c oxidase by reacting with oxidised copper: a general mechanism for the interaction of copper proteins with nitric oxide? *FEBS Lett.* **414:** 281–284.
35. Hagen, T. *et al.* 2003. Redistribution of intracellular oxygen in hypoxia by nitric oxide: effect on HIF1alpha. *Science* **302:** 1975–1978.
36. Gjedde, A. *et al.* 2005. Cerebral metabolic response to low blood flow: possible role of cytochrome oxidase inhibition. *J. Cereb. Blood Flow Metab.* **25:** 1183–1196.
37. Thomas, D.D. *et al.* 2001. The biological lifetime of nitric oxide: implications for the perivascular dynamics of NO and O2. *Proc. Natl. Acad. Sci. USA* **98:** 355–360.
38. Boveris, A. & E. Cadenas. 1975. Mitochondrial production of superoxide anions and its relationship to the antimycin insensitive respiration. *FEBS Lett.* **54:** 311–314.
39. Poderoso, J.J. *et al.* 1996. Nitric oxide inhibits electron transfer and increases superoxide radical production in rat heart mitochondria and submitochondrial particles. *Arch. Biochem. Biophys.* **328:** 85–92.
40. Hollis, V.S. *et al.* 2003. Monitoring cytochrome redox changes in the mitochondria of intact cells using multi-wavelength visible light spectroscopy. *Biochim. Biophys. Acta* **1607:** 191–202.
41. Alvarez, S. *et al.* 2003. Oxygen dependence of mitochondrial nitric oxide synthase activity. *Biochem. Biophys. Res. Commun.* **305:** 771–775.
42. Koppenol, W.H. 2001. 100 years of peroxynitrite chemistry and 11 years of peroxynitrite biochemistry. *Redox Rep.* **6:** 339–341.
43. Brown, G.C. 2007. Nitric oxide and mitochondria. *Front Biosci.* **12:** 1024–1033.
44. Butterfield, D.A. *et al.* 2007. Elevated levels of 3-nitrotyrosine in brain from subjects with amnestic mild cognitive impairment: implications for the role of nitration in the progression of Alzheimer's disease. *Brain Res.* **1148:** 243–248.

45. Zhang, Y.J. *et al.* 2006. Peroxynitrite induces Alzheimer-like tau modifications and accumulation in rat brain and its underlying mechanisms. *FASEB J.* **20:** 1431–1442.
46. Burmester, T. *et al.* 2000. A vertebrate globin expressed in the brain. *Nature* **407:** 520–523.
47. Mammen, P.P. *et al.* 2006. Cytoglobin is a stress-responsive hemoprotein expressed in the developing and adult brain. *J. Histochem. Cytochem.* **54:** 1349–1361.
48. Schelshorn, D.W. *et al.* 2009. Expression of hemoglobin in rodent neurons. *J. Cereb. Blood Flow Metab.* **29:** 585–595.
49. Santos, R.M. *et al.* 2011. Brain nitric oxide inactivation is governed by the vasculature. *Antioxid. Redox Signal.* **14:** 1011–1021.
50. Kesner, R.P., I. Lee & P. Gilbert. 2004. A behavioral assessment of hippocampal function based on a subregional analysis. *Rev. Neurosci.* **15:** 333–351.
51. Garthwaite, J. 2008. Concepts of neural nitric oxide-mediated transmission. *Eur. J. Neurosci.* **27:** 2783–2802.
52. Grivas, I. *et al.* 2003. Vascular network of the rat hippocampus is not homogeneous along the septotemporal axis. *Brain Res.* **971:** 245–249.
53. Fekete, A. *et al.* 2008. Layer-specific differences in reactive oxygen species levels after oxygen-glucose deprivation in acute hippocampal slices. *Free Radic. Biol. Med.* **44:** 1010–1022.
54. Iadecola, C. 1993. Regulation of the cerebral microcirculation during neural activity: is nitric oxide the missing link? *Trends Neurosci.* **16:** 206–214.
55. Girouard, H. & C. Iadecola. 2006. Neurovascular coupling in the normal brain and in hypertension, stroke, and Alzheimer disease. *J. Appl. Physiol.* **100:** 328–335.
56. Lovick, T.A., L.A. Brown & B.J. Key. 1999. Neurovascular relationships in hippocampal slices: physiological and anatomical studies of mechanisms underlying flow-metabolism coupling in intraparenchymal microvessels. *Neuroscience* **92:** 47–60.
57. Meng, W., J.R. Tobin & D.W. Busija. 1995. Glutamate-induced cerebral vasodilation is mediated by nitric oxide through N-methyl-D-aspartate receptors. *Stroke* **26:** 857–862.
58. Faraci, F.M. & K. R. Breese. 1993. Nitric oxide mediates vasodilatation in response to activation of N-methyl-D-aspartate receptors in brain. *Circ. Res.* **72:** 476–480.
59. Pelligrino, D.A. *et al.* 1996. NO synthase inhibition modulates NMDA-induced changes in cerebral blood flow and EEG activity. *Am. J. Physiol.* **271:** H990–H995.
60. Dirnagl, U., U. Lindauer & A. Villringer. 1993. Role of nitric oxide in the coupling of cerebral blood flow to neuronal activation in rats. *Neurosci. Lett.* **149:** 43–46.
61. Akgoren, N., M. Fabricius & M. Lauritzen. 1994. Importance of nitric oxide for local increases of blood flow in rat cerebellar cortex during electrical stimulation. *Proc. Natl. Acad. Sci. USA* **91:** 5903–5907.
62. Iadecola, C. *et al.* 1995. Nitric oxide contributes to functional hyperemia in cerebellar cortex. *Am. J. Physiol.* **268:** R1153–R1162.
63. Adachi, K. *et al.* 1994. Increases in local cerebral blood flow associated with somatosensory activation are not mediated by NO. *Am. J. Physiol.* **267:** H2155–H2162.
64. Greenberg, J.H., N.W. Sohn & P.J. Hand. 1999. Nitric oxide and the cerebral-blood-flow response to somatosensory activation following deafferentation. *Exp. Brain Res.* **129:** 541–550.
65. Wang, Q. *et al.* 1993. Nitric oxide does not act as a mediator coupling cerebral blood flow to neural activity following somatosensory stimuli in rats. *Neurol. Res.* **15:** 33–36.
66. Lindauer, U. *et al.* 1999. Nitric oxide: a modulator, but not a mediator, of neurovascular coupling in rat somatosensory cortex. *Am. J. Physiol.* **277:** H799–H811.
67. Gordon, G.R., S.J. Mulligan & B.A. MacVicar. 2007. Astrocyte control of the cerebrovasculature. *Glia* **55:** 1214–1221.
68. Harder, D.R., C. Zhang & D. Gebremedhin. 2002. Astrocytes function in matching blood flow to metabolic activity. *News Physiol. Sci.* **17:** 27–31.

Ann. N.Y. Acad. Sci. ISSN 0077-8923

ANNALS OF THE NEW YORK ACADEMY OF SCIENCES
Issue: *Environmental Stressors in Biology and Medicine*

Physiological functions of GPx2 and its role in inflammation-triggered carcinogenesis

Regina Brigelius-Flohé and Anna Patricia Kipp

Biochemistry of Micronutrients Department, German Institute of Human Nutrition, Potsdam-Rehbruecke, Germany

Address for correspondence: Dr. Regina Brigelius-Flohé, German Institute of Human Nutrition, Potsdam-Rehbruecke, Arthur-Scheunert-Allee 114–116, D-14558 Nuthetal, Germany. flohe@dife.de

Mammalian glutathione peroxidases (GPxs) are reviewed with emphasis on the role of the gastrointestinal GPx2 in tumorigenesis. GPx2 ranks high in the hierarchy of selenoproteins, corroborating its importance. Colocalization of GPx2 with the Wnt pathway in crypt bases of the intestine and its induction by Wnt signals point to a role in mucosal homeostasis, but GPx2 might also support tumor growth when increased by a dysregulated Wnt pathway. In contrast, the induction of GPx2 by Nrf2 activators and the upregulation of COX2 in cells with a GPx2 knockdown reveal inhibition of inflammation and suggest prevention of inflammation-mediated carcinogenesis. The Janus-faced role of GPx2 has been confirmed in a mouse model of inflammation-associated colon carcinogenesis (AOM/DSS), where GPx2 deletion increased inflammation and consequently tumor development, but decreased tumor size. The model further revealed a GPx2-independent decrease in tumor development by selenium (Se) and detrimental effects of the Nrf2-activator sulforaphane in moderate Se deficiency.

Keywords: glutathione peroxidase-2; colon cancer; inflammation; transcriptional regulation; GPx2 knockout; apoptosis

Introduction

The human genome comprises eight genes encoding the glutathione peroxidases GPx1–8,[1,2] of which 1–4 and 6 are selenoproteins. Moreover, the GPx4 gene is expressed in three distinct forms.[3] GPx1–4 reduce hydroperoxides by means of glutathione (GSH), which thereby becomes oxidized to GSSG and subsequently regenerated by glutathione reductase. The existence of five selenoenzymes with similar reaction specificity raises the question for the physiological meaning of this apparent redundancy. Hydroperoxide removal is considered a major role of GPx1, whereas this is less clear for the others. In fact, GPx1–4 differ from each other in respect to their ranking in the hierarchy of selenoproteins, tissue, subcellular localization, substrate specificity, and transcriptional regulation, which collectively points to distinct physiological roles. Highly specialized functions of GPxs have indeed been reported. GPx1, beyond its reported role of balancing inflammation-[4] and infection-[5] associated oxidative stress, is also implicated in the regulation of glucose homeostasis.[6,7] GPx4 inhibits IL-1–induced NF-κB activation[8] and 12,15-lipoxygenase-mediated apoptosis,[9] its nuclear form is involved in chromatin condensation,[10] and the mitochondrial form is indispensable for mammalian spermatogenesis.[11–13] This article briefly compiles some basic facts regarding properties of the mammalian selenium (Se)-GPxs and then focuses on recent insights into potential physiological and cancer-related roles of GPx2.

The GPx family

Distinct substrate specificities

All glutathione peroxidases reduce H_2O_2 and hydroperoxides from free fatty acids. Only GPx4, the phospholipid hydroperoxide GPx (PHGPx), and some of its nonmammalian congeners,[14] also accept hydroper oxides of complex phospholipids even in membranes.[15] More recently, a specific reaction of GPx4 with 12, 15-lipoxygenase products has been identified.[9] GPx3, the extracellular or plasma GPx is able to reduce complex lipids to a certain extent. Substrate specificity of GPx2 is not quite clear; it appears to be similar to GPx1 with some preference for hydroperoxides of linoleic acid (13-HPODE).[16] Thus, substrate specificity reflects solubility of GPxs, with GPx4 being a more lipophilic enzymes also working in membranes. By the reaction with specific substrate involved in different signaling cascades, GPxs might affect these cascades in a unique manner.

Typically, the reducing substrate of the mammalian Se-GPxs is GSH. A strict GSH specificity has been documented for GPx1 and is likely similar for GPx2. Discrete reactivity with glutaredoxin and thioredoxin, which is the preferred substrate of most non-mammalian (non-Se) GPxs,[14] were reported for GPx3.[17] GPx4 is least specific for GSH and even reacts with SH groups of various proteins, although not promiscuously.[18]

Localization

GPx1 is ubiquitously expressed and mainly localized in the cytoplasm, but also in mitochondria and in some cells also in the peroxisomes (reviewed in Ref. 19). GPx3 is primarily synthesized in the kidney[20] and from there released as glycosylated protein into the extracellular space.[21,22] Also GPx4 is ubiquitously expressed but is especially high in spermatids, while in spermatozoa it is present as an enzymatically inactive structural protein.[12] GPx2 has first been found in the gastrointestinal system, accordingly named GI-GPx or GPx-GI, and suggested to act as a barrier against hydroperoxide absorption.[16,23] It has also been found in epithelial cells of the lung[24] and, even more interestingly, in several cancer cells of epithelial origin.[25–28] GPx2 expression in the intestine is high at crypt bases and gradually declines to the top of the crypts or villi in the colon or small intestine, respectively.[26]

Ranking in the hierarchy of selenoproteins

Selenoproteins are not uniformly supplied with Se. Some disappear fast when Se becomes limiting (low ranking), others keep being synthesized even at low Se availability (high ranking). RNA of low ranking proteins is rapidly degraded in Se deficiency, whereas RNA of high ranking ones remain stable, enabling a faster resynthesis upon reavailability of Se. Within the family of GPxs, the ranking is GPx2 $>$ GPx4 $\gg$ GPx3 $=$ GPx1 established in cell culture.[29] In a study of the mouse colon, only GPx1 and GPx3 RNA was decreased, whereas GPx2 and GPx4 RNA remained stable under marginal Se deficiency.[30] From its high ranking in the hierarchy it has been inferred that the function of GPx2 might be more important than that of other GPxs. However, only GPx4, as opposed to GPx2 and the others, proved to be of vital importance.[31]

Peculiarities of GPx2

Transcriptional regulation

The promoter of GPx2 harbors binding sites for the transcription factors Nrf2, β-catenin/TCF, ΔNp63, Nkx3.1, and the retinoic acid receptor (reviewed in Ref. 32) from which the first three directly activate GPx2 expression. The first transcription factor detected to induce GPx2 was Nrf2.[33] The GPx2 promoter contains two times the respective responsive element, that is, the antioxidant RE (ARE), or better called electrophile RE (EpRE), from which one turned out to be functionally active. The promoter activity was stimulated by Nrf2 activators such as sulforaphane (SFN), curcumin, and *t*-butyl hydrochinone.[33] Dependence of oxidation-mediated GPx2 induction by Nrf2 has also been shown in the lung of wild type (WT) and Nrf-2-knockout mice, thus confirming the regulation *in vivo*.[24,34–36] Since Nrf2 generally upregulates enzymes of the adaptive response and, thus, supports cell survival, a protective role of GPx2 appears plausible at first glance. However, also Nrf2 can be upregulated in certain cancers providing a selective advantage for survival of respective cancer cells.[37–39]

GPx2 is the first selenoprotein identified as target of the Wnt pathway.[40] Its promoter contains five β-catenin/TCF binding sites (TBE), from which one was functionally active. The Wnt pathway controls the expression of genes required for proliferation of stem cells at crypt bases, a fundamental process to maintain mucosal homeostasis. Stem cells differentiate into their final destinations, that is, enterocytes, goblet, and enteroendocrine cells during migration to the top of crypts or villi, whereas when differentiated into Paneth cells, stay at the crypt bottom in the small intestine.[41] Both the colocalization of the Wnt pathway and GPx2, as well as the regulation of GPx2 via Wnt signals, point to a role of GPx2 in the continuous self-renewal of intestinal epithelium and, thus, in the mucosal homeostasis. Since a constitutive activation of the Wnt pathway is characteristic for colon cancer cells, the support of proliferation in these cells by GPx2 might not be rated as particularly anticarcinogenic. A procarcinogenic function of GPx2 can also be deduced from its regulation by ΔNp63,[42] a transcription factor highly expressed in undifferentiated cells. Upregulation of GPx2 by ΔNp63 led to an inhibition of oxidation-induced apoptosis. Regulation of GPx2 by this transcription factor supports the idea that GPx2 is involved in the self-renewal of intestinal mucosa, but also raises scepticism about a clear anticarcinogenic function of GPx2.

The common denominator in the described transcriptional regulation of GPx2 expression might be thiol oxidation. Inflammatory cells release oxidizing compounds like peroxides, leukotrienes or prostaglandins, and proinflammatory mediators, which signal via the production of reactive oxygen species,[43,44] whereas cancer cells need elevated levels of reactive oxygen species for there own survival.[45] In both pathways, redox sensors transduce signals upon oxidation.[46] In the Nrf2 pathway, Keap1 sequesters Nrf2 in the cytosol and prepares it for degradation. Upon oxidation of specific thiols in Keap1, Nrf2 is released and translocated into the nucleus to activate target genes.[47] Similarly, reduced nucleoredoxin (Nrx) blocks disheveled (Dvl), which otherwise would transduce Wnt signals.[48] Upon oxidation of Nrx by NADPH oxidase (Nox) 1-derived H_2O_2,[49] Dvl is released and inhibits glycogen synthase kinase (GSK)-3β in the β-catenin degradation complex. Thereby, β-catenin is stabilized and can activate target genes (Fig. 1). The role of the resulting upregulation of GPx2 is still unclear (see below).

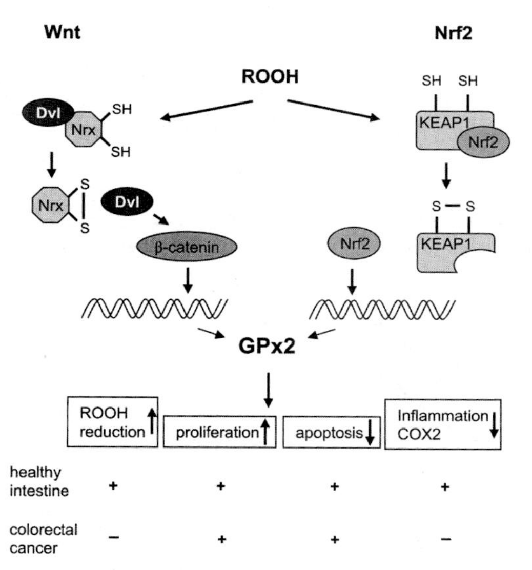

	ROOH reduction ↑	proliferation ↑	apoptosis ↓	Inflammation COX2 ↓
healthy intestine	+	+	+	+
colorectal cancer	–	+	+	–

Figure 1. Oxidation as possible common denominator in the transcriptional regulation of GPx2 expression by the Nrf2 and the Wnt pathway in the intestine. Key steps in the Wnt pathway are the binding of Wnt proteins to the Frizzled receptor, release of disheveled (Dvl) from nucleoredoxin (Nrx), binding of Dvl to the β-catenin degradation complex, thereby inhibiting glycogen synthase kinase 3β as well as the subsequent phosphorylation, ubiquitination, and degradation of β-catenin. β-catenin moves into the nucleus and activates target genes. Nrx binds to Dvl in the reduced form and is released upon oxidation of thiol groups by Nox1-derived H_2O_2.[49] In the Nrf2-pathway Keap1 sequesters Nrf2 in the cytosol and marks it for proteasomal degradation by ubiquitination. The association between Keap1 and Nrf2 is disturbed when certain cysteines in Keap1 are oxidized. Nrf2 can no longer be exposed for ubiquitination; it translocates into the nucleus and activates target genes. The upregulation of GPx2 results in hydroperoxide removal, increased proliferation, inhibition of apoptosis, and a decrease in COX2 expression and inflammation. Whereas all effects are positive for healthy cells, cancer cells will only profit from support for proliferation and from prevention of their elimination by apoptosis. Removal of hydroperoxides and inhibition of inflammation, rather, will inhibit cancer cell growth. By scavenging hydroperoxides, GPx2 might turn off the activity of the Wnt and Nrf2 pathway in a feedback loop.

Deletion of GPx2 in vitro

To learn more about its putative function, GPx2 was stably downregulated in HT29 cells by siRNA. Cells, thus obtained by transfection of two different siRNAs and selection of at least two clones of each siRNA, had lost GPx2 RNA and protein almost completely.[50] At the same time, COX2, microsomal prostaglandin E synthase, and PGE_2 production were highly increased. The siGPx2 cells exhibited an enhanced invasive potential and migrated faster in a wound-healing assay.[51] Both effects obviously required upregulated COX2 activity since celecoxib, a specific COX2 inhibitor, rescued the effects to the level observed in control cells. From these findings, an anti-inflammatory potential of GPx2 could be expected. In contrast, siGPx2 cells were not able to grow anchorage independently and developed into much smaller tumors than WT cells when injected into nude mice.[51] These effects rather point to a support of proliferation not only of healthy intestinal epithelial cells but also of tumor cells by GPx2.

Deletion of GPx2 in vivo

In contrast to GPx1/2 double-knockout mice, which develop colitis and intestinal cancer spontaneously, GPx1 or GPx2 single-knockout mice did not show any obvious phenotype.[52] However, in line with its intestinal localization, a highly increased number of apoptotic cells was found at colonic crypt bases of GPx2 single-knockout (GPx2KO) mice, especially in mice fed a diet moderately deficient in Se.[53] Interestingly, GPx1 was remarkably upregulated in GPx2KO mice in the very same areas even when fed the Se-poor diet. From these observations, an attempt to substitute for GPx2 by upregulation of GPx1 can be suggested. This attempt was only partly successful since, although apoptosis was less pronounced in GPx2KO mice when fed a Se-supranutritional diet, it was not completely prevented. Thus, the role of GPx2 in the intestine appears to be unique and cannot easily be taken over by another selenoprotein even not by a member of the same family.

Se, GPx2, and SFN in a model of inflammation-triggered colon carcinogenesis

With the *in vitro* and xenograft findings in mind, the following questions were tested in *in vivo* studies: Can GPx2 prevent inflammation and subsequently carcinogenesis?, Can Se prevent inflammation and subsequently carcinogenesis via enhancing GPx2 synthesis?, and Does SFN prevent inflammation and subsequently carcinogenesis via upregulation of GPx2?

To answer these questions the azoxymethane (AOM)/dextran sulfate sodium (DSS) mouse model was used. AOM/DSS treatment was combined with feeding diets containing different Se contents, with and without SFN application.[54] WT and GPx2KO mice were fed a Se-poor, -adequate, or -supranutritional diet from weaning until the end of the experiment. After four weeks AOM was applied, followed by 1% DSS in the drinking water for seven days. One group of each Se status was given SFN starting one week before or together with AOM, respectively. Tumor formation was analyzed 12 weeks after AOM and correlated with the severity of colitis three weeks after AOM and morphological changes of the colon at the end of the experiment (Fig. 2). All AOM/DSS-treated mice developed colitis, which was generally more severe in GPx2KO mice than in WT under all Se states

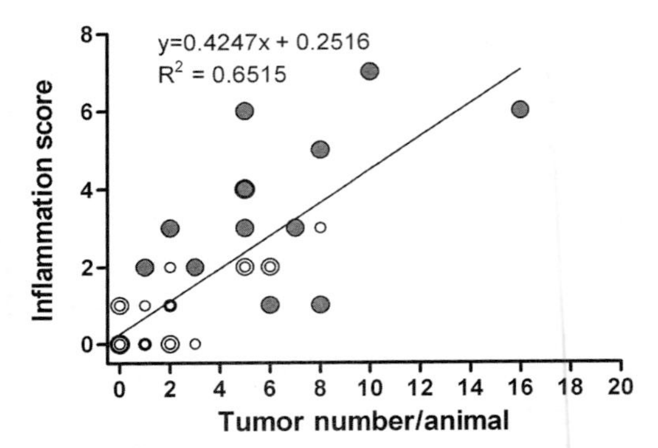

Figure 2. Tumor development correlates with inflammation score. WT and GPx2KO mice adjusted to the three different Se states were treated with AOM/DSS. Clinical inflammation parameters were scored during the acute phase of colitis (during and after DSS application) until signs disappeared (weight loss, occult blood, diarrhea) and at the end of the experiment 12 weeks after AOM injection (morphological changes of the colon such as swelling and shortening). For further details see the text. Linear regression of correlation between inflammation score and tumor number was calculated ($n = 10$ per group). Small empty symbols represent WT, large filled symbols GPx2KO animals. Symbols with a concentric circle indicate identical values for a WT as well as a knockout mouse. Thick borders indicate more than one animal with identical values of the same genotype.

and especially high in moderate Se deficiency. Inflammation, and accordingly tumor formation, was decreased under the Se-supranutritional diet. Tumor numbers/animal tended to be higher in GPx2KO mice at all Se diets and decreased under supranutritional Se. In contrast, tumor size was smaller in Se-poor and Se-supranutritional GPx2KO mice, correlating with the smaller tumors in nude mice developing from HT29 cells in which GPx2 was knocked down (see above).

SFN effects were surprising. SFN applied one week before AOM-enhanced inflammation in Se-poor WT and GPx2KO mice, but decreased inflammation in Se-adequate mice to an identical score in both genotypes. This was not observed in mice fed the Se-supranutritional diet. Similarly, tumor numbers were only diminished by SFN (simultaneously given with AOM) in Se-adequate animals and again were the same in WT and GPx2KO animals.[54] Thus, SFN equalized inflammation and tumor development in GPx2KO and WT under Se-adequacy only, indicating that SFN requires Se for being protective. However, protection cannot be mediated by GPx2 since it was equally observed in GPx2KO mice. A candidate mediator is GPx1, which is upregulated when GPx2 is deleted (see above), but can only be synthesized when Se is available. Upregulation of GPx1 in GPx2KO mice can only partially compensate loss of GPx2 in the inhibition of acute inflammation, which was always more severe in GPx2KO mice at each Se diet. This explains why both enzymes have to be eliminated before spontaneous ileocolitis and cancer can develop.[55]

Taken together, GPx2 appears to be a protective enzyme with anti-inflammatory potential, which might be supported by its antiapoptotic capacity. By inhibition of inflammation tumor formation appears to be diminished in the chosen model. However, inhibition of apoptosis can also explain why tumors from WT cells grow better than tumors from siGPx2 cells in nude mice and in GPx2KO mice. Thus, an anticarcinogenic function of GPx2 appears to depend on the tumor stage[1] and the modulation of tumorigenesis by inflammation. GPx2 might inhibit initiation but—in accordance with its proposed physiological function—support the growth of established tumor cells. Furthermore, the dependence of a preventive effect of SFN on Se highlights the risk of an uncritical intake of glucosinolates if the Se status is low.

Conclusions

Localization of GPx2 at crypt bases and the Wnt pathway suggests a physiological role in the self-renewal of the intestinal epithelium. Apoptosis, which is inhibited by GPx2 in the same areas, may also provide support for that role. Suppression of COX2 expression may indicate anti-inflammatory activity of GPx2, which, along with hydroperoxide removal, might inhibit initiation of carcinogenesis. Inhibition of invasion and migration may then inhibit carcinogenesis at a very late time point (i.e., metastasis). A beneficial role of GPx2 may also be deduced because of its induction by Nrf2 activators. By inhibition of apoptosis and facilitation of cell proliferation, GPx2 might, however, contribute to the support of cancer cell growth. Supporting this line of thinking, the Janus-faced role of GPx2 in tumorigenesis has been confirmed in a mouse model of inflammation-associated carcinogenesis, which addresses the impact of GPx2 genotype, Se status, and the Nrf2 activator SFN. The carcinogenesis model further indicated a GPx2-independent protective effect of Se, likely mediated by GPx1, and the presence of detrimental effects of SFN in moderate Se deficiency, collectively cautioning against an uncritical consumption of Nrf2 activators.

Conflicts of interest

The authors declare no conflicts of interest.

References

1. Brigelius-Flohé, R. & A. Kipp. 2009. Glutathione peroxidases in different stages of carcinogenesis. *Biochim. Biophys. Acta* **1790:** 1555–1568.
2. Nguyen, V.D., M.J. Saaranen, A.R. Karala, *et al.* 2011. Two endoplasmic reticulum PDI peroxidases increase the efficiency of the use of peroxide during disulfide bond formation. *J. Mol. Biol.* **406:** 503–515.
3. Flohé, L. & R. Brigelius-Flohé. 2006. Selenoproteins of the glutathione system. In *Se: Its Molecular Biology and Role in Human Health.* D.L. Hatfield, Eds.: 161–172. Kluwer Academic Publishers. Boston, Dordrecht, London.
4. Jaeschke, H., Y.S. Ho, M.A. Fisher, *et al.* 1999. Glutathione peroxidase-deficient mice are more susceptible to neutrophil-mediated hepatic parenchymal cell injury during endotoxemia: importance of an intracellular oxidant stress. *Hepatology* **29:** 443–450.
5. Beck, M.A., R.S. Esworthy, Y.S. Ho & F.F. Chu. 1998. Glutathione peroxidase protects mice from viral-induced myocarditis. *FASEB J.* **12:** 1143–1149.
6. McClung, J.P., C.A. Roneker, W. Mu, *et al.* 2004. Development of insulin resistance and obesity in mice overexpressing cellular glutathione peroxidase. *Proc. Natl. Acad. Sci. USA* **101:** 8852–8857.
7. Lei, X.G. & M.Z. Vatamaniuk. 2011. Two tales of antioxidant enzymes on beta cells and diabetes. *Antioxid. Redox Signal.* **14:** 489–503.
8. Brigelius-Flohé, R., B. Friedrichs, S. Maurer, *et al.* 1997. Interleukin-1-induced nuclear factor kappa B activation is inhibited by overexpression of phospholipid hydroperoxide glutathione peroxidase in a human endothelial cell line. *Biochem. J.* **328:** 199–203.
9. Seiler, A., M. Schneider, H. Forster, *et al.* 2008. Glutathione peroxidase 4 senses and translates oxidative stress into 12/15-lipoxygenase dependent- and AIF-mediated cell death. *Cell Metab.* **8:** 237–248.
10. Conrad, M., M. Schneider, A. Seiler & G.W. Bornkamm. 2007. Physiological role of phospholipid hydroperoxide glutathione peroxidase in mammals. *Biol. Chem.* **388:** 1019–1025.
11. Ursini, F., S. Heim, M. Kiess, *et al.* 1999. Dual function of the selenoprotein PHGPx during sperm maturation. *Science* **285:** 1393–1396.
12. Flohé, L. 2007. Se in mammalian spermiogenesis. *Biol. Chem.* **388:** 987–995.
13. Conrad, M. 2009. Transgenic mouse models for the vital selenoenzymes cytosolic thioredoxin reductase, mitochondrial thioredoxin reductase and glutathione peroxidase 4. *Biochim. Biophys. Acta* **1790:** 1575–1785.
14. Toppo, S., L. Flohe, F. Ursini, *et al.* 2009. Catalytic mechanisms and specificities of glutathione peroxidases:

variations of a basic scheme. *Biochim. Biophys. Acta* **1790:** 1486–1500.
15. Maiorino, M., J.P. Thomas, A.W. Girotti & F. Ursini. 1991. Reactivity of phospholipid hydroperoxide glutathione peroxidase with membrane and lipoprotein lipid hydroperoxides. *Free Radic. Res. Commun.* **12–13**(Pt 1): 131–135.
16. Wingler, K., C. Müller, K. Schmehl, *et al.* 2000. Gastrointestinal glutathione peroxidase prevents transport of lipid hydroperoxides in CaCo-2 cells. *Gastroenterology* **119:** 420–430.
17. Bjoernstedt, M., J. Xue, W. Huang, *et al.* 1994. The thioredoxin and glutaredoxin systems are efficient electron donors to human plasma glutathione peroxidase. *J. Biol. Chem.* **269:** 29382–29384.
18. Maiorino, M., A. Roveri, L. Benazzi, *et al.* 2005. Functional interaction of phospholipid hydroperoxide glutathione peroxidase with sperm mitochondrion-associated cysteine-rich protein discloses the adjacent cysteine motif as a new substrate of the selenoperoxidase. *J. Biol. Chem.* **280:** 38395–38402.
19. Lubos, E., J. Loscalzo & D.E. Handy. 2011. Glutathione peroxidase-1 in health and disease: from molecular mechanisms to therapeutic opportunities. *Antioxid. Redox Signal.* **15:** 1957–1997.
20. Malinouski, M., S. Kehr, L. Finney, *et al.* 2012. High-resolution imaging of Se in kidneys: a localized Se pool associated with glutathione peroxidase 3. *Antioxid. Redox Signal.* **16:** 185–192.
21. Avissar, N., D.B. Ornt, Y. Yagil, *et al.* 1994. Human kidney proximal tubules are the main source of plasma glutathione peroxidase. *Am. J. Physiol. Gastrointest. Liver Physiol.* **266:** C367–C375.
22. Burk, R.F., G.E. Olson, V.P. Winfrey, *et al.* 2011. Glutathione peroxidase-3 produced by the kidney binds to a population of basement membranes in the gastrointestinal tract and in other tissues. *Am. J. Physiol. Gastrointest. Liver Physiol.* **301:** G32–G38.
23. Chu, F.F., J.H. Doroshow & R.S. Esworthy. 1993. Expression, characterization, and tissue distribution of a new cellular Se-dependent glutathione peroxidase, GSHPx-GI. *J. Biol. Chem.* **268:** 2571–2576.
24. Singh, A., T. Rangasamy, R.K. Thimmulappa, *et al.* 2006. Glutathione peroxidase 2, the major cigarette smoke-inducible isoform of GPX in lungs, is regulated by Nrf2. *Am. J. Respir. Cell Mol. Biol.* **35:** 639–650.
25. Mörk, H., O.H. al-Taie, K. Bahr, *et al.* 2000. Inverse mRNA expression of the selenocysteine-containing proteins GI-GPx and SeP in colorectal adenomas compared with adjacent normal mucosa. *Nutr. Cancer* **37:** 108–116.
26. Florian, S., K. Wingler, K. Schmehl, *et al.* 2001. Cellular and subcellular localization of gastrointestinal glutathione peroxidase in normal and malignant human intestinal tissue. *Free Rad. Res.* **35:** 655–663.
27. Al-Taie, O.H., N. Uceyler, U. Eubner, *et al.* 2004. Expression profiling and genetic alterations of the selenoproteins GI-GPx and SePP in colorectal carcinogenesis. *Nutr. Cancer* **48:** 6–14.
28. Chiu, S.T., F.J. Hsieh, S.W. Chen, *et al.* 2005. Clinicopathologic correlation of up-regulated genes identified using cDNA microarray and real-time reverse transcription-PCR in human colorectal cancer. *Cancer Epidemiol. Biomarkers Prev.* **14:** 437–443.
29. Wingler, K., M. Bocher, L. Flohé, *et al.* 1999. mRNA stability and selenocysteine insertion sequence efficiency rank gastrointestinal glutathione peroxidase high in the hierarchy of selenoproteins. *Eur. J. Biochem.* **259:** 149–157.
30. Kipp, A., A. Banning, E.M. van Schothorst, *et al.* 2009. Four selenoproteins, protein biosynthesis, and Wnt signalling are particularly sensitive to limited Se intake in mouse colon. *Mol. Nutr. Food Res.* **53:** 1561–1572.
31. Imai, H., F. Hirao, T. Sakamoto, *et al.* 2003. Early embryonic lethality caused by targeted disruption of the mouse PHGPx gene. *Biochem. Biophys. Res. Commun.* **305:** 278–286.
32. Banning, A., A. Kipp & R. Brigelius-Flohé. 2012. Glutathione peroxidase 2 and its role in cancer. In *Se: Its Molecular Biology and Role in Human Health.* 3rd ed. D.L. Hatfield, M.J. Berry & V.N. Gladyshev, Eds.: 271–282. Springer Science+Business Media, LLC. New York.
33. Banning, A., S. Deubel, D. Kluth, *et al.* 2005. The GI-GPx gene is a target for Nrf2. *Mol. Cell Biol.* **25:** 4914–4923.
34. Cho, H.Y., A.E. Jedlicka, S.P. Reddy, *et al.* 2002. Role of NRF2 in protection against hyperoxic lung injury in mice. *Am. J. Respir. Cell Mol. Biol.* **26:** 175–182.
35. Cho, H.Y., S.P. Reddy, M. Yamamoto & S.R. Kleeberger. 2004. The transcription factor NRF2 protects against pulmonary fibrosis. *Faseb J.* **18:** 1258–1260.
36. Cho, H.Y., S.P. Reddy, A. Debiase, *et al.* 2005. Gene expression profiling of NRF2-mediated protection against oxidative injury. *Free Radic. Biol. Med.* **38:** 325–343.
37. Lau, A., N.F. Villeneuve, Z. Sun, *et al.* 2008. Dual roles of Nrf2 in cancer. *Pharmacol. Res.* **58:** 262–270.
38. DeNicola, G.M., F.A. Karreth, T.J. Humpton, *et al.* 2011. Oncogene-induced Nrf2 transcription promotes ROS detoxification and tumorigenesis. *Nature* **475:** 106–109.
39. Brigelius-Flohé, R., M. Mueller, D. Lippmann & A.P. Kipp. 2012. The Yin and Yang of Nrf2-regulated selenoproteins in carcinogenesis. *Int. J. Cell Biol.* In press.
40. Kipp, A., A. Banning & R. Brigelius-Flohé. 2007. Activation of the glutathione peroxidase 2 (GPx2) promoter by beta-catenin. *Biol. Chem.* **388:** 1027–1033.
41. Reya, T. & H. Clevers. 2005. Wnt signalling in stem cells and cancer. *Nature* **434:** 843–850.
42. Yan, W. & X. Chen. 2006. GPX2, a direct target of p63, inhibits oxidative stress-induced apoptosis in a p53-dependent manner. *J. Biol. Chem.* **281:** 7856–7862.
43. Federico, A., F. Morgillo, C. Tuccillo, *et al.* 2007. Chronic inflammation and oxidative stress in human carcinogenesis. *Int. J. Cancer* **121:** 2381–2386.
44. Itzkowitz, S.H. & X. Yio. 2004. Inflammation and cancer IV. Colorectal cancer in inflammatory bowel disease: the role of inflammation. *Am. J. Physiol. Gastrointest. Liver Physiol.* **287:** G7–G17.
45. Liou, G.Y. & P. Storz. 2010. Reactive oxygen species in cancer. *Free Radic. Res.* **44:** 479–496.
46. Brigelius-Flohé, R. & L. Flohé. 2011. Basic principles and emerging concepts in the redox control of transcription factors. *Antioxid. Redox Signal.* **15:** 2335–2381.

and, in contrast to messenger RNAs, do not undergo posttranscriptional regulation. Furthermore, the long life span (weeks and months) of proteins reflects the long-term accumulation of phenotype changes. By comparison, transcriptome alterations only reflect short-term changes, due to the short lifetime of RNA molecules (hours). However, the clinical predictivity of proteome analyses is limited by the relatively low number of proteins analyzed. It has been estimated that whole proteomes are composed of 1.5×10^6 proteins.[17] Therefore, only approximately 0.001% of the whole proteome is analyzed when using wide-coverage antibody microarray testing of 500–1,000 proteins. This proportion represents notably poor coverage of the proteome domain, thereby limiting the clinical predictivity and biological relevance of the information provided by proteome analyses.

MicroRNome

MicroRNA represents a major breakthrough in the use of molecular biomarkers to evaluate the effects of chemopreventive agents before cancer onset. The microRNA domain, typically referred to as the miRNome, is currently fully documented. The microRNA domain accounts for approximately 1,200 human microRNAs, all of which have been spotted on available microarray systems, and therefore microRNome (miRNome) coverage is extensive. In contrast to messenger RNA, microRNA does not have the problem of posttranscriptional silencing, and microRNA represents the master posttranscriptional regulator Due to the extensive miRNome coverage and the pivotal biological role of microRNAs, their analysis is expected to have good clinical predictivity, which is relevant for the early evaluation of chemopreventive agents. MicroRNA alterations play a critical role in the pathogenesis of many (perhaps all) cancers,[18] particularly including smoke-induced lung carcinogenesis. Dramatic microRNA alterations accompany and characterize the progressive development of lung cancer, as induced by cigarette smoke (Fig. 2). Cigarette smoke induces early microRNA downregulation,[19] which becomes irreversible after long-term exposure.[20] In human cancer, irreversible microRNA silencing can be caused by the homozygous deletion of microRNA genes. Homozygous deletions on chromosome 3p21.1 are a typical genetic alteration detected in human lung cancer, causing the loss of genes encoding for microRNA let-7.[21] Due to these biological conditions, the modulation of early microRNA alterations by chemopreventive agents is a relevant molecular end point for cancer prevention.

A major problem in cancer chemoprevention is the transferability of data obtained in animal experimental models to humans. MicroRNAs are highly conserved during evolution and exhibit interspecies stability between rodents and humans. Therefore, the transferability of the results obtained for microRNAs in rodents to humans is remarkable. This transferability has been demonstrated by comparing microRNA alterations induced by cigarette smoke in mouse and rat lungs with microRNA alterations detected in lung biopsies of human smokers. These comparative studies detected a significant overlap in smoke-altered microRNA patterns between rodents and humans.[22] In another study, rat and human miRNAs were compared for their response to cigarette smoke in the lungs. Nineteen of the 20 human homologues (95%) responded to environmental cigarette smoke (ECS) in a manner similar to that of their rodent counterparts with a significant correlation between the two species in the variation induced by cigarette smoke ($r = 0.446$, $P < 0.05$).[19]

A further advantage to using microRNA testing for the development of chemopreventive agents is that extracellular microRNAs are released into the blood, which allows them to be collected in humans using noninvasive procedures.[23] This characteristic will make it feasible to use microRNA analysis as an intermediate molecular end point in future human chemoprevention trials using noninvasive sampling.

On the basis of these issues, microarray microRNA analysis has been proposed as a method for evaluating the efficacy and safety of chemopreventive agents before the appearance of cancer.[24] Safety was defined as a lack of alterations induced in miRNomes by chemopreventive agents. Efficacy was defined as the ability of chemopreventive agents to counteract microRNA alterations induced by exposure to carcinogens, such as cigarette smoke. Regarding safety, it should be noted that the same chemopreventive agents that were unable to alter the microRNA profile in the lungs may alter microRNA profiles in other organs. Consideration should be given to the liver because chemopreventive agents are typically administered orally.

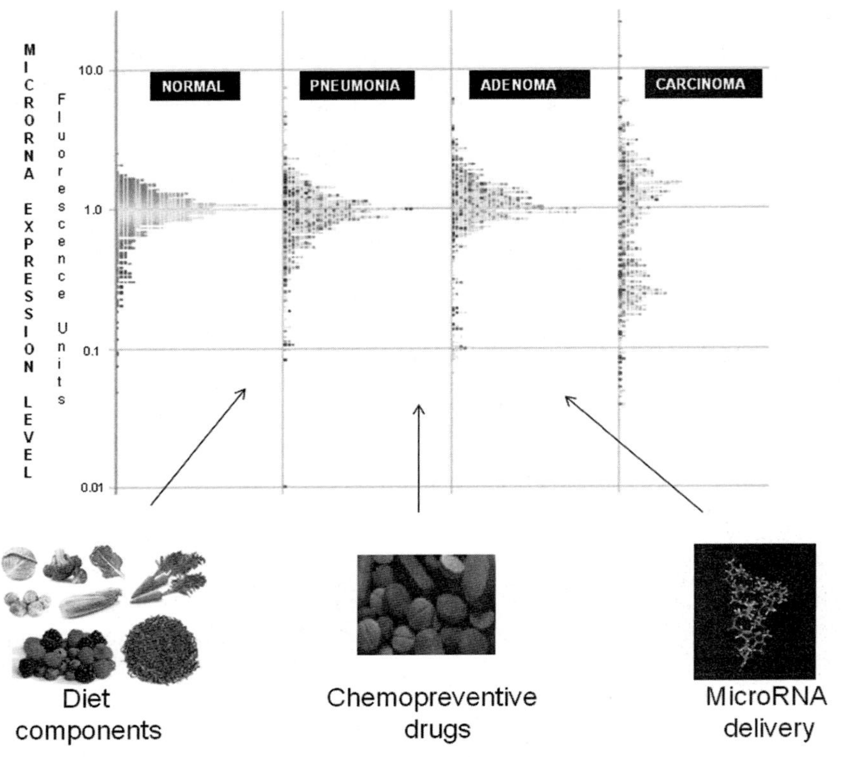

Figure 2. Chemopreventive strategies aimed at modulating microRNA alterations occurring during carcinogenesis before the appearance of cancer. As evaluated by microarray, the microRNA profile is progressively altered in parallel with the histopathological status (normal, pneumonia, adenoma, or carcinoma) in the lungs of mice exposed to cigarette smoke. As a preventive strategy, dietary modulation of the microRNA profile is appropriate during early carcinogenesis phases. Effective chemopreventive drugs should be administered at intermediate phases, and delivery of microRNA may be proposed at advanced phases of carcinogenesis.

Oral administration of the bifunctional inducer phenyl isothiocyanate causes negligible alterations of the microRNA in the lungs while markedly altering the microRNA expression in the livers of mice.[25]

MicroRNA analysis is useful for revealing the molecular mechanisms triggered by chemopreventive agents but is limited by the fact that a single microRNA recognizes multiple target genes. MicroRNA databases (e.g., http://www.targetscan.org/) are currently used to identify microRNA target genes. However, the reliability of bioinformatic data is questionable and should be confirmed by biological analyses.

MicroRNAs as cancer chemopreventive agents

MicroRNAs can be used as passive biomarkers in biomonitoring tools that test the efficacy of chemopreventive drugs in modulating the carcinogenesis process before the onset of cancer. In addition, microRNAs can be employed as active biomarkers because they can be used as chemopreventive agents on their own, thereby exerting an active effect on the modulation of the carcinogenesis process. Given that carcinogens alter microRNA expression, it is conceivable that the normal microRNA profile could be restored before the onset of cancer by administering microRNA, thereby arresting the progression of the cancer. This strategy is expected to normalize the expression of multiple genes, as a single microRNA recognizes multiple target genes. To test the feasibility of this approach, we delivered microRNA let-7a to bronchial epithelial differentiated H727 cells exposed *in vitro* to cigarette smoke condensate. As evaluated by cytofluorometry using a green fluorescent protein (GFP)-labeled let-7a carrier, 26% of the cells were properly transfected. This percentage is relatively high and supports the feasibility of this approach.

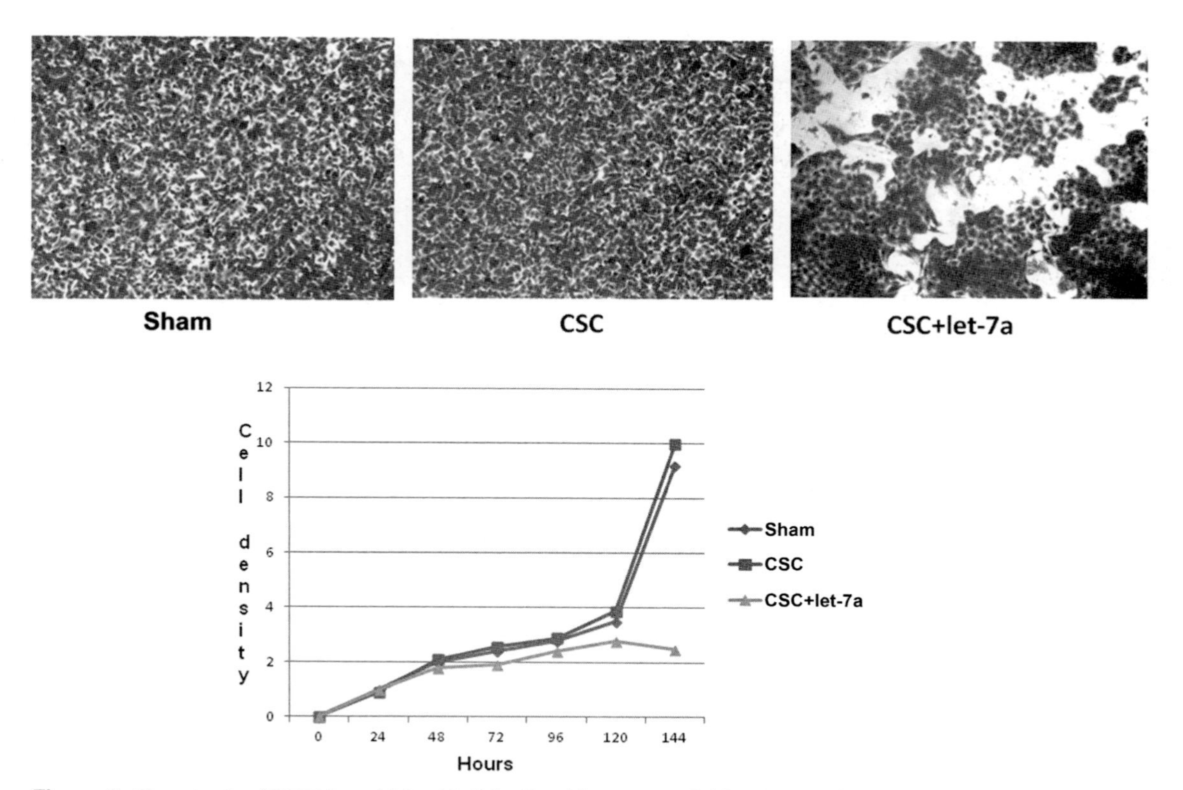

Figure 3. The growth of H727 bronchial epithelial cells, either untreated (Sham), treated with 400 μg/mL cigarette smoke condensate (CSC), or transfected with microRNA let-7a before CSC treatment (CSC + let-7a).

Cigarette smoke typically induces let-7a downregulation,[18] making cells resistant to smoke toxicity by preventing apoptosis. *In vitro*, the number of H727 cells that are resistant to cigarette smoke condensate toxicity is notably high, with 20% of cells surviving at high smoke-condensate doses (400 ug/mL; Fig. 3). Let-7a downregulation allows bronchial epithelial cells to survive and progress throughout the carcinogenesis process. Smoke-induced let-7a downregulation may be corrected *in vitro* by let-7a transfection. In let-7a transfected H727 cells, the number of cells surviving at high condensate doses (400 ug/mL) is reduced to only 3% (Fig. 3). These *in vitro* results indicate that microRNA delivery is a promising preventive strategy to counteract lung carcinogenesis and warrants future studies under *in vivo* conditions.

Conclusions

Molecular medicine makes it feasible to test the safety and efficacy of cancer chemopreventive agents in currently healthy organisms before the onset of cancer. This approach has been applied to experimental *in vivo* animal models as a prelude to the use of molecular intermediate biomarkers in human trials. However, a reliable biomarker requires many characteristics, including high clinical predictivity, relevance of mechanistic and biological information, and the possibility of analysis by noninvasive procedures. All of these characteristics cannot be found in a single biomarker when a complex pathological process, such as carcinogenesis, is examined. Accordingly, future strategies for the use of molecular medicine in the development of cancer chemopreventive agents should use multiple biomarkers that cover the entire process of genetic information from DNA to posttranscriptional regulation to early phenotype alterations. MicroRNA represents a promising molecular intermediate biomarker, the analysis of which provides relevant biological and mechanistic information and may be performed by noninvasive procedures. Furthermore, microRNA modulated by chemopreventive agents may be an effective chemopreventive agent on its own.

Acknowledgment

This study was supported by the Italian Association for Cancer Research Grant Number 8909.

Conflicts of interest

The author declares no conflicts of interest.

References

1. World Cancer Res. Fund/American Inst. Cancer Res. Food. 2007. *Nutrition, Physical Activity and the Prevention of Cancer: A global perspective*. American Inst. Cancer Res. Washington DC.
2. Surh, Y.J. 2003. Cancer chemoprevention with dietary phytochemicals. *Nat. Rev. Cancer* **3:** 768–780.
3. De Flora, S., M. Astengo, D. Serra & C. Bennicelli. 1986. Inhibition of urethan-induced lung tumors in mice by dietary N-acetylcysteine. *Cancer Lett.* **32:** 235–241.
4. van Zandwijk, N., O. Dalesio, U. Pastorino, *et al.* 2000. EUROSCAN, a randomized trial of vitamin A and N-acetylcysteine in patients with head and neck cancer or lung cancer. *J. Natl. Cancer Inst.* **92:** 977–986.
5. Bowen, D.J., M. Thornquist, K. Anderson, *et al.* 2003. Carotene and Retinol Efficacy Trial. Stopping the active intervention: CARET. *Control Clin. Trials* **24:** 39–50.
6. Beta Carotene Cancer Prevention Study Group. 1994. The effect of vitamin E and beta carotene on the incidence of lung cancer and other cancers in male smokers. *N. Eng. J. Med.* **330:** 1029–1035.
7. Burn, J., A.M. Gerdes, F. Macrae, *et al.*; CAPP2 Investigators. 2011. Long-term effect of aspirin on cancer risk in carriers of hereditary colorectal cancer: an analysis from the CAPP2 randomised controlled trial. *Lancet* **378:** 2081–2087.
8. Izzotti, A., R.M. Balansky, F. D'Agostini, *et al.* 2001. Modulation of biomarkers by chemopreventive agents in smoke exposed rats. *Cancer Res.* **61:** 2472–2479.
9. Van Schooten, F.J., A. Besaratinia, S. De Flora, *et al.* 2002. Effects of oral administration of N-acetyl-L-cysteine: a multibiomarker study in smokers. *Cancer Epidemiol. Biomarkers Prev.* **11:** 167–175.
10. Shen, J., M.D. Gammon, M.B. Terry, *et al.* 2005. Polymorphisms in XRCC1 modify the association between polycyclic aromatic hydrocarbon-DNA adducts, cigarette smoking, dietary antioxidants, and breast cancer risk. *Cancer Epidemiol. Biomarkers Prev.* **14:** 336–342.
11. Lee, M.S., L. Su & D.C. Christiani. 2010. Synergistic effects of NAT2 slow and GSTM1 null genotypes on carcinogen DNA damage in the lung. *Cancer Epidemiol. Biomarkers Prev.* **19:** 1492–1497.
12. Izzotti, A., A. Camoirano, C. Cartiglia, *et al.* 1999. Patterns of DNA adduct formation in liver and mammary epithelial cells of rats treated with 7,12-dimethylbenz(a)anthracene, and selective effects of chemopreventive agents. *Cancer Res.* **59:** 4285–4290.
13. Phillips, D.H. 2005. DNA adducts as markers of exposure and risk. *Mutat. Res.* **577:** 284–292.
14. Izzotti, A., M. Bagnasco, C. Cartiglia, *et al.* 2005. Modulation of multigene expression and proteome profiles by chemopreventive agents. *Mutat. Res.* **591:** 212–223.
15. Izzotti, A., M. Bagnasco, C. Cartiglia, *et al.* 2004. Proteomic analysis as related to transcriptome data in the lung of chromium(VI) treated rats. *Int. J. Oncol.* **24:** 1513–1522.
16. Izzotti, A., M. Bagnasco, C. Cartiglia, *et al.* 2005. Chemoprevention of genome, transcriptome, and proteome alterations induced by cigarette smoke in rat lung. *Eur. J. Cancer* **41:** 1864–1874.
17. Melton, L. 2004. Proteomics in multiplex. *Nature* **429:** 101–107.
18. Calin, G.A. & C.M. Croce. 2006. MicroRNA-cancer connection: the beginning if a new tale. *Cancer Res.* **66:** 7390–7394.
19. Izzotti, A., G. Calin, P. Arrigo, *et al.* 2009. Downregulation of microRNA expression in the lung of rats exposed to cigarette smoke. *FASEB J.* **23:** 806–812.
20. Izzotti, A., C. Cartiglia, M. Longobardi, *et al.* 2011. Dose responsiveness and persistence of microRNA alterations induced by cigarette smoke in mice. *Mutat. Res.* **717:** 9–16.
21. Calin, G.A., C. Sevignani, C.D. Dumitru, *et al.* 2004. Human microRNA genes are frequently located at fragile sites and genomic regions involved in cancers. *Proc. Natl. Acad. Sci. USA* **101:** 2999–3004.
22. Perdomo, C., A. Spira & F. Schembri. 2011. MiRNAs as regulators of the response to inhaled environmental toxins and airway carcinogenesis. *Mutat. Res.* **717:** 32–37.
23. Etheridge, A., I. Lee, L. Hood, *et al.* 2011. Extracellular microRNA: a new source of biomarkers. *Mutat. Res.* **717:** 85–90.
24. Izzotti, A., G. Calin, V.E. Steele, *et al.* 2010. Chemoprevention of cigarette smoke induced alterations of microRNA expression in rat lung. *Cancer Prev. Res.* **3:** 62–72.
25. Izzotti, A., P. Larghero, C. Cartiglia, *et al.* 2010. Modulation of microRNA expression by budesonide, phenethyl isothiocyanate, and cigarette smoke in mouse liver and lung. *Carcinogenesis* **31:** 894–901.

Ann. N.Y. Acad. Sci. ISSN 0077-8923

ANNALS OF THE NEW YORK ACADEMY OF SCIENCES
Issue: *Environmental Stressors in Biology and Medicine*

Experimental basis for discriminating between thermal and athermal effects of water-filtered infrared A irradiation

Tobias Jung and Tilman Grune

Institute of Nutrition, Friedrich Schiller University Jena, Jena, Germany

Address for correspondence: Tilman Grune, Institute of Nutrition, Friedrich Schiller University Jena, Dornburger Strasse 24, 07745 Jena, Germany. tilman.grune@uni-jena.de

Considering the widespread application of water-filtered infrared A (wIRA) irradiation in medicine, cosmetics, and wellness, we have conluded that the biological effects of this electromagnetic spectrum, ranging from 780 nm to 1400 nm, have become an important focus of experimental research. Two main effects of wIRA on single cells are discussed: thermal effects, caused by absorption of energy by cellular water and the aqueous medium surrounding the irradiated sample that result in warming, and supposed athermal effects that result from a direct interaction of wIRA with cellular molecules/structures excluding water. In the following, we discuss different experimental setups and highlight some cellular responses to thermal and athermal wIRA effects, as well as the experimental problems in differentiating between them.

Keywords: water-filtered infrared A (wIRA); single cells; thermal effects; athermal effects; water cooling; air cooling

Introduction

Infrared A, covering the solar spectra from 780 nm to 1400 nm, represents about 30% of the whole-solar irradiance, which is composed of the UV, visible, and infrared spectra. After significant atmospheric filtering by water, oxygen, and CO_2, the spectral irradiance of the solar irradiation reaching the earth's surface is massively reduced—ranging from 920 to 980, 1100 to 1180, and 1360 to 1430 nanometers. In clinical applications, filtering of the wavelengths that are mainly absorbed by water is achieved by using a cuvette containing a water layer of 4, 7, or 10 mm, while 7 mm of water is the most often used filter. After water filtering, higher irradiance levels are applicable, and combined with higher exposure times, the dehydration of skin/tissue is reduced, the heat sensitivity during irradiation decreases, and the penetration depth increases up to 5 centimeters. The known immediate effects of water-filtered infrared A (wIRA) application are an increase in skin/tissue temperature, of the oxygen partial pressure in the exposed tissue, and of tissue perfusion.

One of the most important factors is the sterile application of wIRA, rendering it ideal for treatment of (chronic) open wounds, acute operation wounds, chronic leg ulcer, burns, and application of photodynamic therapy. Previously, warmth was applied to chronic wounds (like ulcers of the lower leg) via a water bath. Naturally, such a treatment holds a massive risk of infection because of direct contact of the surface of the wound with water. In contrast, application of wIRA can be made without direct contact of any device,[1] clearly reducing any risk of wound contamination. wIRA effects are especially ascribed to the wavelengths from 800 to 900,[2–4] 800,[5] 820,[6–8] and 830 nm.[9]

Beyond the essential medical applications for prevention and therapy, the use of wIRA exposure to skin, body parts, or even the whole body has become widespread in cosmetics and wellness. In light of this, wIRA-induced effects have become a focus of interest in experimental research.

Thermal and athermal effects of wIRA

One key aspect of the current discussion about the biological effects of wIRA is the induction of thermal and athermal cellular responses triggered by wIRA exposure and how to differentiate between

doi: 10.1111/j.1749-6632.2012.06581.x

both effects. Since pure athermal effects, that is, effects of wIRA that are not mediated exclusively by warming of the irradiated sample, are especially detectable in single cells, this review will center on this topic. In larger tissue samples, built of many single cells or even layers of different cell types, warming of the sample due to wIRA absorption cannot be avoided or reliably excluded by an effective cooling setup. Thus, it is very difficult in this scenario to investigate athermal effects without any influence of thermal effects.

There is no doubt about the existence of thermal effects that primarily result from wIRA absorption in cellular water, which contributes to warming of the sample. This is significantly reduced, but not completely avoided, by water filtering of the IRA source. Since water is the most abundant compound of most mammalian cells, the main contribution of warming in the irradiated sample is wIRA absorption. Thermal effects of wIRA can be reproduced by pure warming of the irradiated sample to the steady-state temperature, which would have been otherwise reached by wIRA exposure over the same period of time. According to our experiments, thermal effects of wIRA, which are in fact induced by a poor experimental setup that causes a significant warming of the sample during wIRA exposure, can be reproduced by pure warming of the samples to the same temperature induced by wIRA irradiation under light protection in a water bath. Using this methodology, we were able to detect an increased amount of free radical formation (via DCF-oxidation in living cells), massive increases both in protein oxidation,[10] and mitochondrial superoxide release (data not published), as well as decreases in mitochondrial membrane potential and an increase in the cellular calcium concentration. These cellular responses were identified both after wIRA irradiation without proper cooling and after warming of the samples (single cells) to the average temperature found during wIRA exposure for the same time period.

Athermal effects of wIRA irradiation contribute to cellular responses through a direct interaction or absorption of wIRA by cellular components/structures, with the exception of water; these findings are, however, still the point of considerable debate. One clue to the existence of athermal effects is the direct absorption of wIRA by compounds of the mitochondrial respiratory chain, as found by Karu *et al.* According to their results, cytochrome c oxidase (complex IV) exhibited absorption maxima at 767, 791, and 880 nm, where 767 nm is not in the wIRA range. These peaks are produced by absorption in both the heme structures and the two copper atoms of complex IV. The two atoms (Cu_A and Cu_B) can be found either in an oxidized (ox) or reduced (red) state, overall constituting four possible states with four different absorption peaks: 620 ($Cu_{A(red)}$), 680 ($Cu_{B(ox)}$), 760 ($Cu_{B(red)}$), and 820 nm ($Cu_{A(ox)}$).[11,12] However, only ($Cu_{A(ox)}$) absorbs in the wIRA range. Furthermore, suspension of isolated mitochondria shows only very low wIRA absorption compared to the isolated cytosolic aqueous fraction of human dermal fibroblasts.[10] In addition to the respiratory chain's complex IV, other

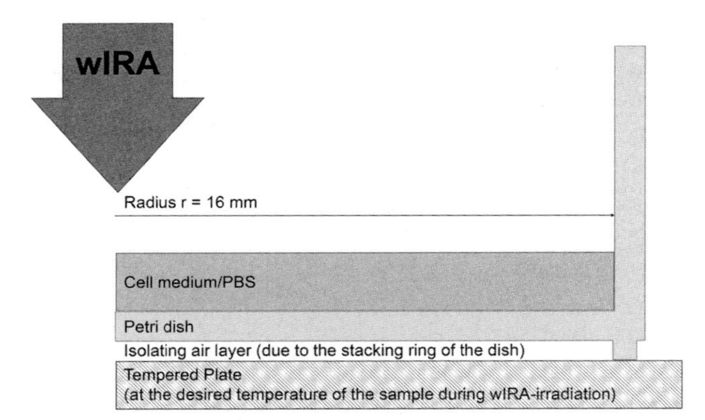

Figure 1. Cross-section of a commonly used air-cooling setup. This panel shows a schematic cross-section of a commonly used air-cooling setup. In contrast to the water-cooling setup shown in Figure 2, heat is removed passively. Please note the two layers of isolating material (polystyrene and air) between the cooling plate (tempered at the desired temperature of the wIRA-irradiated sample) and the sample itself (cells at the bottom of the Petri dish, not shown).

proteins in cells or in the tissue may be able to absorb parts of the wIRA spectra as they contain copper ions surrounded by a similar molecular environment (aromatic carbon rings), such as different matrix metalloproteinases or the several present superoxide dismutases (SODs).

Other abundant cellular components, which are able to absorb in the wIRA range, are proteins (at 910 and 1020 nm), lipids (920 and 1040 nm), and glucose (939 and 1126 nm).[13] In the current literature, it is assumed that in mammalian cells, athermal effects of wIRA result from the induction of the different matrix metalloproteinases, such as MMP-1,[14–18] -2,[19] -3,[20] -9,[15] and -13.[20] Activation of TGF-β1[19] and the MAPK pathways have been observed, in addition to: release of cytochrome c and Smac/DIABLO[21] from the mitochondria, induction of the translocation of Bax from the cytosol into the nucleus,[21] an increase of the heat shock proteins HSP27[21] and HSP70,[22] and a general increase in the cellular formation of reactive oxygen species (ROS).[10,18] Furthermore, decreases in cellular carotenoid concentration[23] were found as well as an induction of both trypase[24] and p53.[25]

In contrast, other experimental outcomes did not suggest possible athermal effects of wIRA irradiation. Applegate *et al.* reported a lack of induction of antioxidative proteins, such as heme oxygenase (HO)-1, SOD, different heat shock proteins (HSPs), proteases, NO-synthases, or increased damage of DNA, both in human skin cells and skin.[26] Additionally, other work reports no changes in HSP70-[27,28] or MMP-1 expression,[29] AF or procaspase induction,[21] DNA damage,[25] cell survival,[16,29] or induction of ROS-formation[10] after biological samples are exposed to wIRA.

Piazena and Kelleher have recently published a comprehensive review of cellular responses after wIRA exposure in single cells and skin. [30] Unfortunately, as the authors note, much of the methodological description in such studies is insufficient or absent. This includes information about the experimental setups for wIRA exposure (e.g., wIRA emitter, water filter thickness, other filter sets, the resulting spectral irradiance of the unit); the method

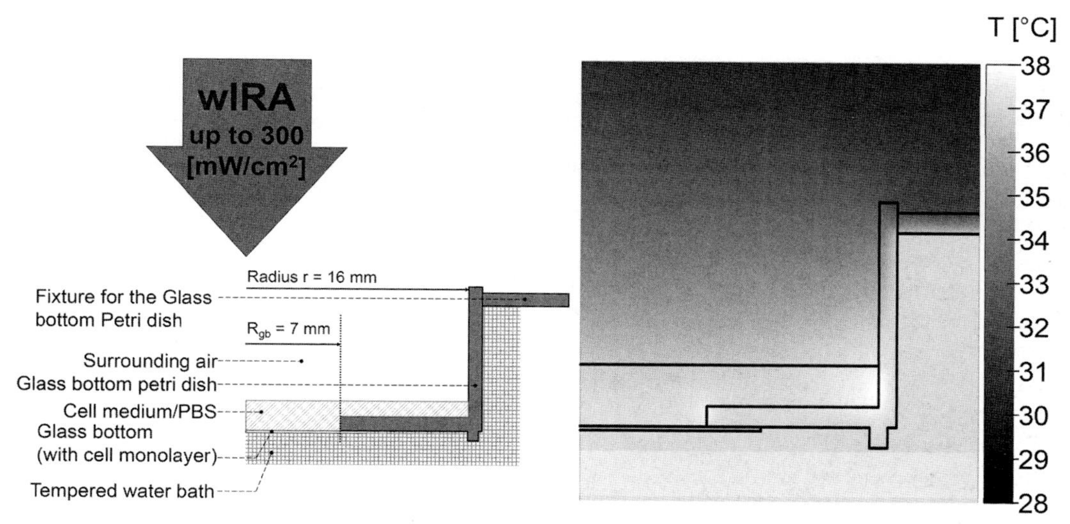

Figure 2. A realistic computer simulation of the temperature distribution found in a water-cooled glass-bottom dish during an applied wIRA irradiance of 240 mW/cm^2. In the left panel, a cross-section of the water-cooling setup, used for both experimental wIRA exposure and computer simulation, is depicted. The right panel shows the steady-state temperature distribution calculated by a computer simulation of the setup during wIRA irradiation (wIRA-irradiance of 240 mW/cm^2). The computed distribution was experimentally verified. The water temperature was constant at 37 °C, and the air temperature was set at 25 °C. Considering the negligible volume and heat capacity of a single cell on the glass bottom compared to the surrounding medium, we have assumed the temperature of the cells to be the same as the temperature of their environment. The maximal difference between the desired temperature of the sample cells during wIRA exposure (determined by the cooling water) was in the range of +0.1 °C at irradiances of up to 300 mW/cm^2.

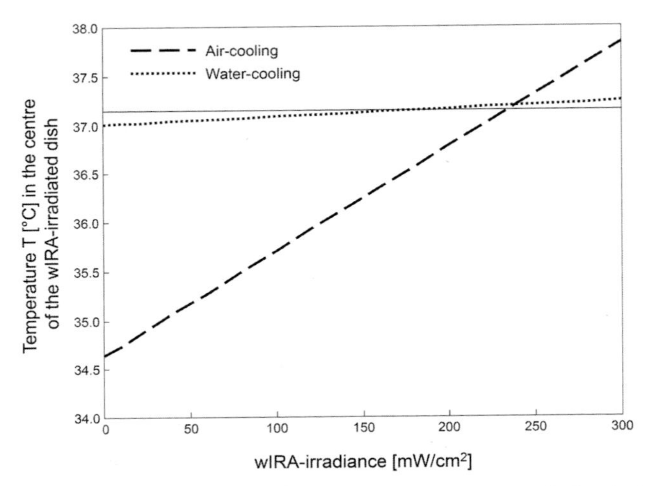

Figure 3. Steady-state sample temperature of air and water cooling dependent on the applied wIRA irradiance. This figure shows the steady-state temperatures of sample cells dependent on the applied wIRA irradiance for both air (dashed line) and water cooling (dotted line) according to the experimental setups described above. The temperature was directly measured via microthermometry in the center of the irradiated Petri dish.

of cooling or temperature regulation of the sample during irradiation (i.e., air or watercooling, temperature measurement of the sample); environmental conditions; room temperature and humidity; the absorption spectra of the media used and cell culture materials; or the thermography employed in the experimental setup.

Experimental methods for the detection of athermal wIRA effects

It is important to note that when only those with sufficient methodological detail are included, the majority of wIRA studies report athermal effects from air cooling. In contrast, those studies using a water-cooling method did not report these effects.

Further, measurement of the temperature at the direct site of the irradiated sample cells was only done in a few studies. In the following sections, we will provide and overview and a discussion of the two most frequently used experimental setups that provide temperature homeostasis of wIRA-irradiated samples during exposure.

Air cooling

In several publications the use of air cooling is described—typically, the sample to be irradiated is placed atop a tempered plate with a Petri dish. A representative scheme of such a setup is shown in Figure 1. Some of the main problems that occur with this setup are the very low heat conduction of polystyrene (the common cell culture material), which prevents effective cooling of the irradiated sample, in addition to an isolating air layer (due to the stacking ring of the dish) of up to 1 mm in thickness between the Petri dish and the tempered plate. Furthermore, in contrast to water cooling (as described below), heat is only passively and not actively removed from the system by forced convection. Another factor influencing the steady-state temperature of the wIRA-exposed sample is the wIRA interaction with the cooling-plate itself; if the sample exhibits high absorption, the sample temperature can significantly differ from the plate's temperature. Another factor that shows very strong influence is the room temperature: an increase or decrease of 5 °C of room temperature may increase or decrease the steady-state temperature of the irradiated sample by about 1 °C,[31] which may be sufficient to trigger a significant cellular response to hyperthermia or even heat shock. At low wIRA irradiances, the sample temperature is primarily determined by the room temperature, not by the temperature of the plate.

Water cooling

An effective cooling setup is depicted in Figure 2 and described in detail in another publication by our group.[31] In the experimental setup, the constant temperature of the cooling water is achieved via a tempered, stirred water bath with a minimal capacity of about 20 L, while cooling water is (7 L/min) continually washed around the glass bottom that holds the sample cells. The use of water has the advantage of a significantly higher heat capacity (4180 J/g∗K) compared to air (1005 J/g∗K), enabling a higher deduction of thermal energy from the wIRA-irradiated system per unit of time. The glass bottom is approximately 0.14 mm thick and has better heat conductivity (1.16 W/m∗K) than the common cell culture material polystyrene (0.17 W/m∗K), which in most dishes is approximately 0.8–1.0 mm in thickness. The efficiency of the sample temperature homeostasis increases with the diameter of the glass bottom. Such a cooling setup is able to keep samples at a desired temperature, which is adjusted via the cooling water temperature at irradiances applied up to 300 mW/cm^2 (see Fig. 3).

Comparison of air and water cooling

In Figure 3, the efficiencies of both air and water cooling are compared. Using a water-cooling setup, the maximal deviation of the sample versus the cooling water temperature can be controlled for up to an irradiance of about 300 mW/cm^2.[31] In contrast, the sample temperature in the case of air cooling is mainly determined by the room temperature and the wIRA absorption of the plate used. Thus, air cooling allows for larger differences between the desired compared with the achieved temperature of the sample.

Conclusion

Investigations of athermal effects of wIRA require a diligent and reliable sample temperature homeostasis that is only possible using water cooling as it is described above. If a clear differentiation between thermal and athermal wIRA effects is not possible due to (even slight) warming of the sample, it is not possible to conclude whether any athermal effects were produced. Importantly, temperature homeostasis via water cooling is only effective in the investigation of single cells. Large cell clusters of, for example, skin samples show a much lower surface to volume ratio, thus a warming cannot be prevented as effectively.

Acknowledgments

We thank Dr. Helmut Franz Piazena for detailed physical advice during fine tuning of the computer simulations, Dr. Eduard Wolf for his technical support, and the "Dr. med. h.c. Erwin Braun Stiftung, Basel" for support of our theoretical and experimental work.

Conflicts of interest

The authors declare no conflicts of interest.

References

1. Mercer, J.B., S.P. Nielsen & G. Hoffmann. 2008. Improvement of wound healing by water-filtered infrared-A (wIRA) in patients with chronic venous stasis ulcers of the lower legs including evaluation using infrared thermography. *Ger. Med. Sci.* **6:** Doc11.
2. Albrecht-Buehler, G. 1991. Surface extensions of 3T3 cells towards distant infrared light sources. *J. Cell Biol.* **114:** 493–502.
3. Albrecht-Buehler, G. 1994. Cellular infrared detector appears to be contained in the centrosome. *Cell Motil. Cytoskeleton* **27:** 262–271.
4. Albrecht-Buehler, G. 2005. A long-range attraction between aggregating 3T3 cells mediated by near-infrared light scattering. *Proc. Natl. Acad. Sci. USA* **102:** 5050–5055.
5. Ehrlicher, A., T. Betz, B. Stuhrmann, *et al.* 2002. Guiding neuronal growth with light. *Proc. Natl. Acad. Sci. USA* **99:** 16024–16028.
6. Karu, T.I., L.V. Pyatibrat & G.S. Kalendo. 2001. Cell attachment modulation by radiation from a pulsed light diode (lambda = 820 nm) and various chemicals. *Lasers Surg. Med.* **28:** 227–236.
7. Karu, T.I., L.V. Pyatibrat & G.S. Kalendo. 2001. Cell attachment to extracellular matrices is modulated by pulsed radiation at 820 nm and chemicals that modify the activity of enzymes in the plasma membrane. *Lasers Surg. Med.* **29:** 274–281.
8. Karu, T.I., L.V. Pyatibrat & G.S. Kalendo. 2001. Donors of NO and pulsed radiation at lambda = 820 nm exert effects on cell attachment to extracellular matrices. *Toxicol. Lett.* **121:** 57–61.
9. Chow, R.T., G.Z. Heller & L. Barnsley. 2006. The effect of 300 mW, 830 nm laser on chronic neck pain: a double-blind, randomized, placebo-controlled study. *Pain* **124:** 201–210.
10. Jung T., A. Hohn, H. Piazena & T. Grune. 2010. Effects of water-filtered infrared A irradiation on human fibroblasts. *Free Radic. Biol. Med.* **48:** 153–160.
11. Karu, T.I., L.V. Pyatibrat, S.F. Kolyakov & N.I. Afanasyeva. 2005. Absorption measurements of a cell monolayer relevant to phototherapy: reduction of cytochrome c oxidase under near IR radiation. *J. Photochem. Photobiol. B.* **81:** 98–106.
12. Karu T. 1999. Primary and secondary mechanisms of action of visible to near-IR radiation on cells. *J. Photochem. Photobiol. B.* **49:** 1–17.
13. Khalil, M.M., M.M. Aboaly & R.M. Ramadan. 2005. Spectroscopic and electrochemical studies of ruthenium and osmium complexes of salicylideneimine-2-thiophenol Schiff base. *Spectrochim. Acta A Mol. Biomol. Spectrosc.* **61:** 157–161.

14. Kim, M.S., Y.K. Kim, K.H. Cho & J.H. Chung. 2006. Regulation of type I procollagen and MMP-1 expression after single or repeated exposure to infrared radiation in human skin. *Mech. Ageing Dev.* **127:** 875–882.
15. Cho, S., M.J. Lee, M.S. Kim, *et al.* 2008. Infrared plus visible light and heat from natural sunlight participate in the expression of MMPs and type I procollagen as well as infiltration of inflammatory cell in human skin in vivo. *J. Dermatol. Sci.* **50:** 123–133.
16. Buechner N., P. Schroeder, S. Jakob, *et al.* 2008. Changes of MMP-1 and collagen type Ialpha1 by UVA, UVB and IRA are differentially regulated by Trx-1. *Exp. Gerontol.* **43:** 633–637.
17. Schroeder P., J. Lademann, M.E. Darvin, *et al.* 2008. Infrared radiation-induced matrix metalloproteinase in human skin: implications for protection. *J. Invest. Dermatol.* **128:** 2491–2497.
18. Schroeder P., C. Pohl, C. Calles, *et al.* 2007. Cellular response to infrared radiation involves retrograde mitochondrial signaling. *Free Radic. Biol. Med.* **43:** 128–135.
19. Danno K., N. Mori, K. Toda, *et al.* 2001. Near-infrared irradiation stimulates cutaneous wound repair: laboratory experiments on possible mechanisms. *Photodermatol. Photoimmunol. Photomed.* **17:** 261–265.
20. Kim, H.H., M.J. Lee, S.R. Lee, *et al.* 2005. Augmentation of UV-induced skin wrinkling by infrared irradiation in hairless mice. *Mech. Ageing Dev.* **126:** 1170–1177.
21. Frank, S., L. Oliver, C.C. Lebreton-De, *et al.* 2004. Infrared radiation affects the mitochondrial pathway of apoptosis in human fibroblasts. *J. Invest. Dermatol.* **123:** 823–831.
22. Kim, M.S., Y.K. Kim, K.H. Cho, J.H. Chung. 2006. Infrared exposure induces an angiogenic switch in human skin that is partially mediated by heat. *Br. J. Dermatol.* **155:** 1131–1138.
23. Darvin, M.E., A. Patzelt, M. Meinke, *et al.* 2009. Influence of two different IR radiators on the oxidative potential of the human skin. *Laser Phys. Lett.* **6:** 229–234.
24. Kim, M.S., Y.K. Kim, D.H. Lee, *et al.* 2009. Acute exposure of human skin to ultraviolet or infrared radiation or heat stimuli increases mast cell numbers and tryptase expression in human skin in vivo. *Br. J. Dermatol.* **160:** 393–402.
25. Frank S., S. Menezes, C.C. Lebreton-De, *et al.* 2006. Infrared radiation induces the p53 signaling pathway: role in infrared prevention of ultraviolet B toxicity. *Exp. Dermatol.* **15:** 130–137.
26. Applegate, L.A., C. Scaletta, R. Panizzon, *et al.* 2000. Induction of the putative protective protein ferritin by infrared radiation: implications in skin repair. *Int. J. Mol. Med.* **5:** 247–2451.
27. Schieke S., H. Stege, V. Kurten, *et al.* 2002. Infrared-A radiation-induced matrix metalloproteinase 1 expression is mediated through extracellular signal-regulated kinase 1/2 activation in human dermal fibroblasts. *J. Invest. Dermatol.* **119:** 1323–1329.
28. Jantschitsch C., S. Majewski, A. Maeda, *et al.* 2009. Infrared radiation confers resistance to UV-induced apoptosis via reduction of DNA damage and upregulation of antiapoptotic proteins. *J. Invest. Dermatol.* **129:** 1271–1279.
29. Gebbers N., N. Hirt-Burri, C. Scaletta, *et al.* 2007. Water-filtered infrared-A radiation (wIRA) is not implicated in cellular degeneration of human skin. *Ger. Med. Sci.* **5:** Doc08.
30. Piazena H. & D.K. Kelleher. 2010. Effects of infrared-A irradiation on skin: discrepancies in published data highlight the need for an exact consideration of physical and photobiological laws and appropriate experimental settings. *Photochem. Photobiol.* **86:** 687–705.
31. Jung T., A. Hohn, A.M. Lau, *et al.* 2012. An experimental setup for the measurement of nonthermal effects during water-filtered infrared a-irradiation of Mammalian cell cultures. *Photochem. Photobiol.* **88:** 371–380.

Ann. N.Y. Acad. Sci. ISSN 0077-8923

ANNALS OF THE NEW YORK ACADEMY OF SCIENCES
Issue: *Environmental Stressors in Biology and Medicine*

Acrolein effects in pulmonary cells: relevance to chronic obstructive pulmonary disease

Nadia Moretto, Giorgia Volpi, Fiorella Pastore, and Fabrizio Facchinetti
Department of Pharmacology, Chiesi Farmaceutici SpA, Parma, Italy

Address for correspondence: F. Facchinetti, Department of Pharmacology, Chiesi Farmaceutici SpA, Via Palermo 26/A, 43122, Parma, Italy. f.facchinetti@chiesi.com

Acrolein (2-propenal) is a highly reactive α,β-unsaturated aldehyde and a respiratory irritant that is ubiquitously present in the environment but that can also be generated endogenously at sites of inflammation. Acrolein is abundant in tobacco smoke, which is the major environmental risk factor for chronic obstructive pulmonary disease (COPD), and elevated levels of acrolein are found in the lung fluids of COPD patients. Its high electrophilicity makes acrolein notorious for its facile reaction with biological nucleophiles, leading to the modification of proteins and DNA and depletion of antioxidant defenses. As a consequence, acrolein results in oxidative stress as well as altered intracellular signaling and gene transcription/translation. In pulmonary cells, acrolein, at subtoxic concentrations, can activate intracellular stress kinases, alter the production of inflammatory mediators and proteases, modify innate immune response, induce mucus hypersecretion, and damage airway epithelium. A better comprehension of the mechanisms underlying acrolein effects in the airways may suggest novel treatment strategies in COPD.

Keywords: acrolein; stress kinases; inflammation; glutathione; proteases; oxidative stress; neutrophils

Introduction

Chronic obstructive pulmonary disease (COPD) is a respiratory disorder characterized by progressive, not fully reversible, airflow limitation associated with an abnormal pulmonary inflammatory response to noxious particles or gases.[1] The most common symptoms of COPD include chronic bronchitis, abnormal sputum production, emphysema, wheeze, and dyspnea with exertion.[2] Despite the fact that COPD is a leading cause of morbidity and mortality worldwide, there are no therapeutic interventions capable of arresting the progression of the disease.[3]

Acrolein (2-propenal) is abundant among the numerous chemical components (>4,700) of cigarette smoke, which is accepted as the most important risk factor for COPD.[4] Acrolein is a highly electrophilic α,β-unsaturated aldehyde, volatile at room temperature, highly irritant to the airways,[5] and toxic. Even if ubiquitously present in the environment,[6] acrolein exposure through cigarette smoke is generally considered to make up a large proportion of total human exposure. In addition, acrolein can be generated endogenously during inflammation, through the degradation of threonine into acrolein by myeloperoxidase (MPO), which is released by neutrophils at the site of inflammation (see Fig. 1).[7] Another endogenous source of acrolein is from the catabolism of polyamines, such as spermine and spermidine, through an amine oxidase–mediated mechanism.[7] Augmented levels of acrolein are found in the expired breath condensate (45.3 nM) and induced sputum of patients with COPD (131.1 nM).[8]

Acrolein interaction with biomolecular targets

Being a water-soluble α,β-unsaturated aldehyde, acrolein possesses a C-3 carbon that can react with nucleophilic sites in proteins, i.e., the sulfhydryl group of cysteine, the imidazole moiety of histidine, the ε-amino group of lysine, and the guanidine group of arginine, thereby impairing protein function.[7] Preferentially acrolein forms

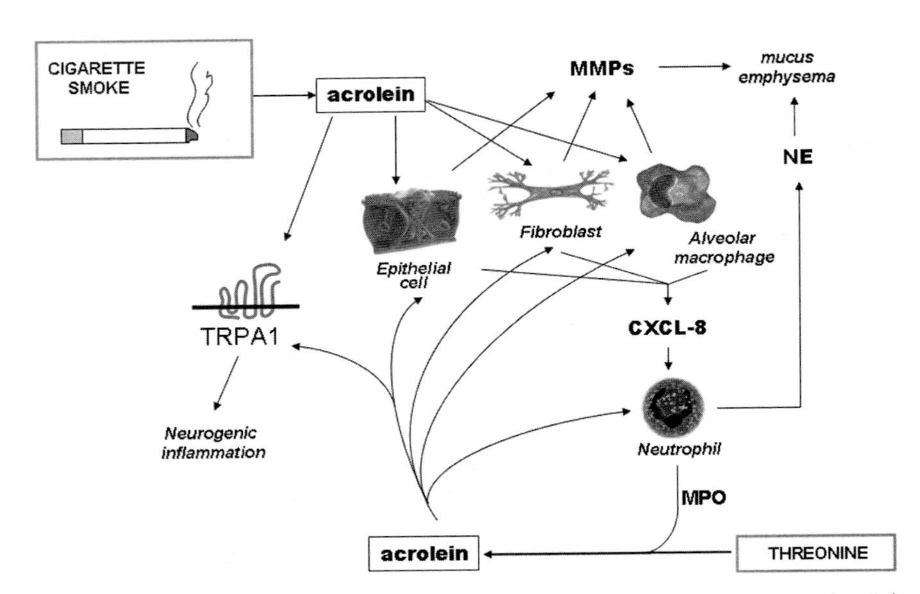

Figure 1. Cigarette smoke contains acrolein, which can stimulate the release of metalloproteases (MMPs) and CXCL-8 (interleukin-8) from alveolar macrophages, epithelial cells, and fibroblasts. CXCL-8 induces recruitment and activation of neutrophils and consequent release of myeloperoxidase (MPO), which causes the formation of acrolein from threonine. Endogenously generated acrolein further stimulates CXCL-8, NE (neutrophilic elastase) and MMPs release and, through TRPA1 activation, can trigger neurogenic inflammatory responses.

Michael-type adducts on cysteine residues.[9] Acrolein exposure was reported to lead to a dose-dependent loss of cellular glutathione (GSH) upon conjugation and formation of GSH–acrolein covalent adducts. GSH is a critical endogenous antioxidant, abundantly present at millimolar concentrations in most cell compartments, such as in the cytosol (1–11 mM), nuclei (3–15 mM), and mithocondria (5–11 mM).[10] GSH plays a key role in cellular reductive processes and detoxification of harmful oxidative species and various xenobiotics.[11] Although acrolein, at levels found in COPD lungs, cannot cause a substantial depletion of intracellular antioxidant defenses (i.e., GSH),[5] it forms adducts also with other key proteins involved in redox signaling, including thioredoxins (Trxs), peroxiredoxins (Prxs), thioredoxin reductase, and GSH S-transferase π.[12] Acrolein adducts affect the activity of thioredoxin reductase (TrxR) and the redox status of Trxs and Prxs in human bronchial epithelial cells.[13] In addition, acrolein is capable of forming adducts, which alter apoptosis, inflammation, mucociliar transport, alveolar–capillary barrier integrity, as well as many other signaling events[14] independent of the lock-and-key mode of activation. By forming acrolein–DNA adducts with deoxyguanosine moiety of DNA, acrolein may also affect DNA replication, transcription, and repair capacity.[15]

Acrolein targets structural macromolecules, forming adducts and cross-links with cytoskeletal targets, i.e., actin and vimentin.[16,17] Acrolein can alter upstream and/or downstream components of a signaling cascade, including transcription factors such as NF-κB,[18,19] which is critically involved in the expression of more than 400 genes regulating antioxidant defense, apoptosis, inflammatory, and immunological responses. Acrolein exposure forms adducts with a member of the basic leucin zipper transcription factor family, the nuclear erythroid-2 related factors 2 (Nrf2), which activate the antioxidant response element (ARE), a transcriptional control element that mediates expression of a set of antioxidant proteins and defense enzymes (e.g., heme oxygenase-1, superoxide dismutase, thioredoxin, thioredoxin reductase, GSH reductase, and GSH peroxidise).[20] On a genome-wide basis, acrolein exposure was reported to decrease the metallothionein class of cytoprotective metal-chelating proteins and to induce a persistent upregulation of several members of the early growth response (EGR) class of zinc finger transcription factors, known to be involved in differentiation and mitogenesis.[14]

Acrolein-mediated lung cell toxicity

The mode of cell death induced by acrolein is dose- and cell type–dependent and is influenced by experimental conditions, as extensively reviewed by Bein and Leukauf.[5] In general, acrolein concentrations exceeding 50 μM cause massive disruption to the structure and function of several cell types,[21] resulting mainly in necrosis.[22,23] At concentrations less than 50 μM, pro-apoptotic mechanisms prevail as an increased apoptosis has been observed in several cell types, such as alveolar macrophages,[22] alveolar type II cell-derived cell line (A549),[24] and bronchial epithelial cells.[25] Acrolein-induced apoptosis appears to be mediated by overproduction of reactive oxygen species, mitochondrial membrane hyperpolarization,[26] and activation of p53, a transcription factor and tumor suppressor, involved in regulating the death receptor Fas pathway.[27] At lower concentrations of acrolein (<15 μM in A549), anti-apoptotic processes dominate, in part, through activation of the survival factor AKT.[26]

Acrolein and airway inflammation

Cigarette smoke exposure clearly affects the immune system, compromising the host ability to mount proper immune and anti-inflammatory responses and contributing to smoking-related pathologies.[28] As a consequence, smoking and COPD are related to bacterial colonization of the pulmonary tract, and rising airway bacterial load is associated with greater airway inflammation and an accelerated lung functional decline.[29] Similar to cigarette smoke, acrolein can cause impairment of pulmonary anti-bacterial defenses.[30,31] In particular, acrolein is known to interfere with innate macrophagic functions, including adhesion, phagocytosis, and homodimerization of Toll-like receptor 4.[32–34] Recently, molecular mechanisms underlying acrolein ability to modulate immune responses have started to be elucidated. Acrolein is known to activate multiple stress kinase pathways such as extracellular-signal-regulated-kinase (ERK1/2) and p38 MAPK, which modulate expression and release of inflammatory mediators, such as CXCL-8/IL-8, tumor necrosis factor-α (TNF-α), and vascular endothelial growth factor (VEGF).[35–37] In addition, acrolein can suppress innate macrophage responses by direct alkylation of c-Jun N-terminal kinase[34] and induce COX-2 in rat lung epithelial cells through the ERK1/2 pathway.[38]

Acrolein has the potential to regulate transcription of a variety of genes primarily by covalently binding with key transcription factors, including NF-κB, which mediates a major signaling pathway associated with inflammation and cytokine gene expression.[19,39] Acrolein, by alkylating cysteine and arginine residues in the NF-κB–binding domain, can suppress NF-κB signaling[40,41] and, consequently, the release of inflammatory mediators, such as IL-1b, whose expression is predominatly regulated through the NF-κB pathway.[42]

It should also be taken into consideration that diverse cell types may respond differently to acrolein as, for instance, in human macrophagic cells, where TNF-α release is actually stimulated by acrolein.[35] The observation that exposure to acrolein can induce neutrophil influx in the lung of small rodents[43] is in line with data showing that acrolein elicits expression and release of the neutrophil chemoattractant CXCL-8 in a variety of lung cells through a p38 MAPK– and ERK1/2-dependent mechanism.[35,36] In addition, acrolein is capable of inhibiting apoptotic signaling in neutrophils resulting in delayed necrotic cell death.[44,45] This mechanism could lead to decreased neutrophil clearance during the inflammation processes, thus prolonging and enhancing neutrophilic inflammation (see Fig. 1). A recent study[46] identified acrolein as the main agent in cigarette smoke mediating airway neurogenic inflammation via stimulation of transient receptor potential ankyrin subfamily member 1 (TRPA1), a non-selective noxious cold-activated cation channel, suggesting a further role for acrolein in promoting inflammatory conditions in the airways (see Fig. 1).

In sum, acrolein-induced suppression of the host response is accompanied by increased neutrophil chemoattraction and decreased neutrophil clearance, suggesting a link between acrolein and smoking-related suppression of host defense as well as neutrophil infiltration, both events typically associated with COPD.

Acrolein and mucus hypersecretion

In COPD patients, airway inflammation causes an excessive mucus production and decrease in mucociliary clearance resulting in mucus retention and airway obstruction.[47,48]

Mucins are complex glycoproteins that provide the viscoelastic properties of mucus that is essential for the protection of the airways. Cigarette smoke can directly increase the transcription levels of mucin genes, mainly MUC5AC and MUC7,[49] and, to a lesser extent, MUC1 and MUC2,[50] and can also synergistically amplify the responses to proinflammatory cytokines and bacterial infections[51] frequently present in lung of COPD.

Animals exposed to acrolein develop airway epithelial damage, bronchiolitis,[52,53] mucus cell metaplasia, and increased MUC5AC expression,[47] accompanied by excessive macrophage accumulation in the lung.[54,55] As observed in lung and bronchial epithelial cells, acrolein can directly induce excessive mucus production.[53,56] Nevertheless, as shown for cigarette smoke, acrolein may also indirectly induce excessive and persistent mucus production through a mechanism involving matrix metalloproteinases (MMPs) activation,[57] pro-EGF ligand shedding, and EGFR-mediated activation of ERK1/2,[58] JNK,[58] and p38 MAPK[59] signaling. Acrolein, at concentrations similar to those found in COPD sputum, is capable of promoting MMP9 activation through the removal of the inhibitory N-terminal pro-domain from pro-MMP9 or by directly adducting cysteine 319 in the MMP14, an activator of MMP9. Activated MMP9 can proteolitically processes various pro-EGFR ligands, thus inducing an autocrine/paracrine EGFR-MAPK–dependent upregulation of MMP9 and MMP14 that, once released from the cell, can be activated by acrolein or other MMPs.[58–61] Thus, acrolein, by priming the activation of MMP9, stimulates the EGFR-MAPK signaling that, in turn, could contribute to MMPs upregulation and persistent EGFR-dependent mucin production. It has been recently shown that the EGFR/ERK1/2-mediated induction of goblet cell metaplasia and mucin hyperproduction after exposure to acrolein may be a consequence of Ras GTPase activation.[62]

In sum, acrolein stimulatory effects on mucins gene expression and mucin production may play a role in promoting chronic mucus hypersecretion, a pathognomonic feature of COPD.

Acrolein and emphysema

Protease–anti-protease imbalance has been proposed to be the principal cause of tissue destructive process leading to emphysema in COPD.[63] Initially put forward for the elastase activity associated with the cigarette smoke–mediated recruitment of neutrophils,[64] the proteases–anti-proteases hypothesis has been reformulated to include the MMPs as the proteolytic enzymes mainly involved in the pathogenesis of emphysema. Recent studies suggest that the MMPs predominantly involved in the lung tissue destructive processes in response to cigarette smoke exposure as well as in COPD appear to be MMP1, -2, -8, -9, -12, and -14.[65–67]

Several lines of evidence indicate that exposure of cells and animals to acrolein can drive an unbalanced protease activation through EGFR-, MAPKs-, and mTOR-mediated up- or downregulation of certain MMPs and protease inhibitors expression, as demonstrated for MMP1,[68] -9,[69,70] -14,[58] and TIMP3.[57,68] Moreover, acrolein exposure can induce macrophage[71] and neutrophil accumulation[54] with consequent production of high levels of proinflammatory cytokines/chemokines and proteases, including neutrophil elastase, which may contribute to alveolar destruction, tissue injury, and remodeling in COPD (see Fig. 1).[52,71]

Targeting acrolein in COPD therapy

A potential approach for the treatment of COPD could be aimed at neutralizing acrolein. Conjugation of acrolein to GSH is an important mechanism for the detoxification of acrolein. However, circulating GSH cannot be increased to a clinically beneficial extent by oral administration, even with high doses (up to 3 g) of GSH, because the limited systemic availability of oral GSH and the rapid pharmacokinetic.[72] Alternatively, increased GSH level can be obtained by administering its precursors *N*-acetylcysteine (NAC), an acetyl derivative of the amino-acid cysteine, which is also a strong reducing agent. NAC is rapidly metabolized to cysteine, whose free thiol-group has antioxidant and reducing properties, including α,β-unsaturated aldehydes scavenging, and that is a direct precursor of GSH. Because of its ability to reduce disulphide bounds, NAC is largely used as a mucolytic agent.[73] In addition, NAC has the potential to interact directly with oxidants such as α,β-unsaturated aldehydes by forming covalent adducts (Michael adducts).[74] Experimental evidence demonstrates the scavenging effect of NAC against cigarette smoke– and acrolein-induced inflammation and emphysema and some beneficial effects in COPD patients in small clinical studies.[35,36,69,75–78] However, an extensive multicenter study (BRONCUS) reported that at least the dose of NAC 600 mg daily is ineffective in preventing the deterioration in lung function and prevention of exacerbation in patients with COPD followed for three years.[79] Given its limited oral bioavailability and short half life,[80] it is conceivable that higher doses of NAC should be utilized in order to obtain efficacious levels in the lung.[81,82]

Nitrogen-containing carbonyl compounds, such as the anti-hypertensive hydralazine,[83] the classic carbonyl reagent methoxyamine,[84] and the glycoxidation inhibitors, aminoguanidine, and carnosine,[85] are known for their aldehyde-scavenging properties and for their ability to counteract acrolein-induced protein carbonylation. However, nitrogen-containing compounds display slower carbonyl-sequestering kinetics when compared with sulfur-containing scavengers.[86]

Recent studies indicate that a decrease in Nrf2 signaling in patients with COPD may hamper their ability to defend against oxidative stress and electrophiles agents. A strategy for increasing GSH levels and cytoprotective pathways can be accomplished through sulforaphane, an organosulfur compound capable of inducing resistance to acrolein through an Nrf2-dependent mechanism.[87] Analogously, the natural polyphenol resveratrol is capable of inducing GSH synthesis *via* activation of Nrf2 and can protect against cigarette smoke–mediated oxidative stress in human lung epithelial cells.[88] Also the natural phenol curcumin is an inducer of cytoprotective pathways against electrophile agents.[89] However, the "drugability" of many natural polyphenols is hampered by the limited bioavailability *in vivo* and/or rapid inactivation in the gastrointestinal tract.

In sum, many "antioxidant" compounds have the potential to protect cells from the toxic effects of acrolein through either direct scavenging or the induction of cytoprotective pathways that can neutralize acrolein as well as other α,β-unsaturated aldehydes. Although several pharmacological interventions are effective in experimental models, it should be considered that the anatomic lesions in COPD, in particular emphysema, can hardly be reproduced in animal models. Indeed, only with chronic smoke exposure (six months) is there evidence of mild emphysema and pulmonar lesions, which, however, never reach a stage comparable to GOLD III/IV in small rodents.[90–92] At present, NAC is the only "anti-acrolein agent" that has been extensively studied in COPD patients. It is likely that erdosteine, a thiol-containing molecule now currently used in the therapy of COPD as mucolytic and whose active metabolite contains a sulfhydryl group,[73] may also function as acrolein scavenger.

Concluding remarks

Chronic cigarette smoke inhalation is a major risk factor for COPD. Some of the pathological changes induced by cigarette smoke exposure can be replicated by its component acrolein in experimental *in vitro* and *in vivo* models. While acute exposure to high doses of acrolein causes acute lung injury, repeated exposure to lower doses represents a risk factor for development of chronic pulmonary inflammation, reduction of host defense, neutrophil inflammation, mucus hypersecretion, and protease-mediated lung tissue damage. A better comprehension of the effects of acrolein in pulmonary cells may shed light on the mechanisms underlying COPD pathogenesis and pave the way for novel therapeutic approaches.

Conflicts of interest

The authors declare no conflicts of interest.

References

1. Celli, B.R. & W. MacNee. 2004. ATS/ERS Task Force. Standards for the diagnosis and treatment of patients with COPD: a summary of the ATS/ERS position paper. *Eur. Respir. J.* **23:** 932–946.
2. Rabe, K.F. *et al.* 2007. Global strategy for the diagnosis, management, and prevention of chronic obstructive pulmonary disease: GOLD executive summary. Global Initiative for Chronic Obstructive Lung Disease. *Am. J. Respir. Crit. Care Med.* **176:** 532–555.
3. Barnes, P.J. 2007. Chronic obstructive pulmonary disease: a growing but neglected global epidemic. *PLoS Med.* **4:** e112.
4. Stedman, R.L. 1968. The chemical composition of tobacco and tobacco smoke. *Chem. Rev.* **68:** 153–207.
5. Bein, K. & G.D. Leikauf. 2011. Acrolein—a pulmonary hazard. *Mol. Nutr. Food Res.* **55:** 1342–1360.
6. Esterbauer, H., R.J. Schaur & H. Zollner. 1991. Chemistry and biochemistry of 4-hydroxynonenal, malonaldehyde and related aldehydes. *Free Radic. Biol. Med.* **11:** 81–128
7. Stevens, J.F. & C.S. Maier. 2008. Acrolein: sources, metabolism, and biomolecular interactions relevant to human health and disease. *Mol. Nutr. Food Res.* **52:** 7–25.
8. Corradi, M. *et al.* 2004. Comparison between exhaled and sputum oxidative stress biomarkers in chronic airway inflammation. *Eur. Respir. J.* **24:** 1011–1017.

9. Witz, G. 1989. Biological interactions of alpha,beta-unsaturated aldehydes. *Free Radic. Biol. Med.* **7:** 333–349.
10. Valko, M. *et al.* 2007. Free radicals and antioxidants in normal physiological functions and human disease. *Int. J. Biochem. Cell Biol.* **39:** 44–84.
11. Meister, A. & M.E. Anderson. 1983. Glutathione. *Annu. Rev. Biochem.* **52:** 711–760.
12. Spiess, P.C. *et al.* 2011. Proteomic profiling of acrolein adducts in human lung epithelial cells. *J. Proteomics.* **74:** 2380–2394.
13. Myers, C. R. & J.M. Myers. 2009. The effects of acrolein on peroxiredoxins, thioredoxins, and thioredoxin reductase in human bronchial epithelial cells. *Toxicology* **257:** 95–104.
14. Thompson, C.A. & P.C. Burcham. 2008. Genome-wide transcriptional responses to acrolein. *Chem. Res. Toxicol.* **21:** 2245–2256.
15. Liu, X.Y., M.X. Zhu & J.P. Xie. 2010. Mutagenicity of acrolein and acrolein-induced DNA adducts. *Toxicol. Mech. Methods* **20:** 36–44.
16. Dalle-Donne, I. *et al.* 2007. Actin Cys374 as a nucleophilic target of alpha,beta-unsaturated aldehydes. *Free Radic. Biol. Med.* **42:** 583–598.
17. Burcham, P.C., A. Raso & C.A. Thompson. 2010. Intermediate filament carbonylation during acute acrolein toxicity in A549 lung cells: functional consequences, chaperone redistribution, and protection by bisulfite. *Antioxid. Redox Signal.* **12:** 337–347.
18. Horton, N.D. *et al.* 1999. Acrolein causes inhibitor kappaB-independent decreases in nuclear factor kappaB activation in human lung adenocarcinoma (A549) cells. *J. Biol. Chem.* **274:** 9200–9206.
19. Lambert, C. *et al.* 2007. Acrolein inhibits cytokine gene expression by alkylating cysteine and arginine residues in the NF-kappaB1 DNA binding domain. *J. Biol. Chem.*, **282:** 19666–19675.
20. Tirumalai, R. *et al.* 2002. Acrolein causes transcriptional induction of phase II genes by activation of Nrf2 in human lung type II epithelial (A549) cells. *Toxicol. Lett.* **132:** 27–36.
21. Kehrer, J.P. & S.S. Biswal. 2000. The molecular effects of acrolein. *Toxicol. Sci.* **57:** 6–15.
22. Li, L. *et al.* 1997. Acrolein induced cell death in human alveolar macrophages. *Toxicol. Appl. Pharmacol.* **145:** 331–339.
23. Tanel, A. & D.A. Averill-Bates. 2005. The aldehyde acrolein induces apoptosis via activation of the mitochondrial pathway. *Biochim. Biophys. Acta.* **1743:** 255–267.
24. Hoshino, Y. *et al.* 2001. Cytotoxic effects of cigarette smoke extract on an alveolar type II cell derived cell line. *Am. J. Physiol. Lung. Cell Mol. Physiol.* **281:** L509–L516.
25. Nardini, M. *et al.* 2002. Acrolein-induced cytotoxicity in cultured human bronchial epithelial cells. Modulation by alpha-tocopherol and ascorbic acid. *Toxicology* **170:** 173–185.
26. Roy, J. *et al.* 2009. Acrolein induces a cellular stress response and triggers mitochondrial apoptosis in A549 cells. *Chem. Biol. Interact.* **181:** 154–167.
27. Roy, J. *et al.* 2010. Acrolein induces apoptosis through the death receptor pathway in A549 lung cells: role of p53. *Can. J. Physiol. Pharmacol.* **88:** 353–368.
28. Stämpfli, M.R. & G.P Anderson. 2009. How cigarette smoke skews immune responses to promote infection, lung disease and cancer. *Nat. Rev. Immunol.* **9:** 377–384.
29. Schäfer, H. & S. Ewig. 2000. Acute exacerbations in chronic obstructive pulmonary disease (COPD)–microbial patterns and risk factors. *Monaldi. Arch. Chest Dis.* **55:** 415–419.
30. Jakab, G.J. 1977. Adverse effect of a cigarette smoke component, acrolein, on pulmonary antibacterial defenses and on viral-bacterial interactions in the lung. *Am. Rev. Respir. Dis.* **115:** 33–38.
31. Li, L. & A. Holian. 1998. Acrolein: a respiratory toxin that suppresses pulmonary host defense. *Rev. Environ. Health* **13:** 99–108.
32. Low, E.S., R.B. Low & G.M. Green. 1977. Correlated effects of cigarette smoke components on alveolar macrophage adenosine triphosphatase activity and phagocytosis. *Am. Rev. Respir. Dis.* **115:** 963–970.
33. Lee, J.S. *et al.* 2008. Acrolein with an alpha, beta-unsaturated carbonyl group inhibits LPS-induced homodimerization of toll-like receptor 4. *Mol. Cells* **25:** 253–257.
34. Hristova, M. *et al.* 2011. The tobacco smoke component acrolein suppresses innate macrophage responses by direct alkylation of c-Jun-N-terminal kinase. *Am. J. Respir. Cell Mol. Biol.* Jul **21.**
35. Facchinetti, F. *et al.* 2007. Alpha,beta-unsaturated aldehydes in cigarette smoke release inflammatory mediators from human macrophages. *Am. J. Respir. Cell. Mol. Biol.* **37:** 617–623.
36. Moretto, N. *et al.* 2009. alpha,beta-Unsaturated aldehydes contained in cigarette smoke elicit IL-8 release in pulmonary cells through mitogen-activated protein kinases. *Am. J. Physiol. Lung. Cell Mol. Physiol.* **296:** L839–L848.
37. Volpi, G. *et al.* 2011. Cigarette smoke and α,β-unsaturated aldehydes elicit VEGF release through the p38 MAPK pathway in human airway smooth muscle cells and lung fibroblasts. *Br. J. Pharmacol.* **163:** 649–661
38. Sarkar, P. & B.E Hayes. 2007. Induction of COX-2 by acrolein in rat lung epithelial cells. *Mol. Cell Biochem.* **301:** 191–199.
39. Li, L., R.F. Jr Hamilton & A. Holian. 1999. Effect of acrolein on human alveolar macrophage NF-kappaB activity. *Am. J. Physiol.* **277**(3 Pt 1): L550–L557.
40. Ji, C., K.R. Kozak & L.J. Marnett. 2001. IkappaB kinase, a molecular target for inhibition by 4-hydroxy-2-nonenal. *J. Biol. Chem.* **276:** 18223–18228.
41. Valacchi, G. *et al.* 2005. Inhibition of NFkappaB activation and IL-8 expression in human bronchial epithelial cells by acrolein. *Antioxid. Redox. Signal.* **7:** 25–31.
42. Kent, L. *et al.* 2008. Cigarette smoke extract induced cytokine and chemokine gene expression changes in COPD macrophages. *Cytokine.* **42:** 205–216.
43. Kasahara, D.I. *et al.* 2008. Acrolein inhalation suppresses lipopolysaccharide-induced inflammatory cytokine production but does not affect acute airways neutrophilia. *J Immunol.* **181:** 736–745
44. Finkelstein, E.I., M. Cardini & A. van der Vliet. 2001. Inhibition of neutrophil apoptosis by acrolein: a mechanism of tobacco-related lung disease? *Am. J. Physiol. Lung. Cell Mol. Physiol.* **281:** L732–L739.

45. Finkelstein, E.I. *et al.* 2005. Regulation of constitutive neutrophil apoptosis by the alpha,beta-unsaturated aldehydes acrolein and 4-hydroxynonenal. *Am. J. Physiol. Lung. Cell Mol. Physiol.* **289:** L1019–L101928.
46. Andrè, E. *et al.* 2008. Cigarette smoke-induced neurogenic inflammation is mediated by alpha,beta-unsaturated aldehydes and the TRPA1 receptor in rodents. *J. Clin. Invest.* **118:** 2574–2582.
47. Leikauf, G.D. *et al.* 2002. Mucin apoprotein expression in COPD. *Chest* **121**(5 Suppl): 166S–182S.
48. Johnson, D.C. 2011. Airway mucus function and dysfunction. *N. Engl. J. Med.* **364:** 978.
49. Fan, H. & L.A. Bobek. 2010. Regulation of human MUC7 mucin gene expression by cigarette smoke extract or cigarette smoke and pseudomonas aeruginosa lipopolysaccharide in human airway epithelial cells and in MUC7 transgenic mice. *Open Respir. Med. J.* **14:** 63–70.
50. Thai, P. *et al.* 2008. Regulation of airway mucin gene expression. *Annu. Rev. Physiol.* **70:** 405–429.
51. Baginski, T.K. *et al.* 2006. Cigarette smoke synergistically enhances respiratory mucin induction by proinflammatory stimuli. *Am. J. Respir. Cell Mol. Biol.* **35:** 165–174.
52. Costa, D.L. *et al.* 1986. Altered lung function and structure in the rat after subchronic exposure to acrolein. *Am. Rev. Respir. Dis.* **133:** 286–291.
53. Borchers, M.T., S.E. Wert & G.D. Leikauf. 1998. Acrolein-induced MUC5ac expression in rat airways. *Am. J. Physiol.* **274**(4 Pt 1): L573–L581.
54. Borchers, M.T. *et al.* 1999. Monocyte inflammation augments acrolein-induced Muc5ac expression in mouse lung. *Am. J. Physiol.* **277**(3 Pt 1): L489– L497.
55. Dorman, D.C. *et al.* 2008. Respiratory tract responses in male rats following subchronic acrolein inhalation. *Inhal. Toxicol.* **20:** 205–216.
56. Borchers, MT., M.P Carty & G.D. Leikauf. 1999. Regulation of human airway mucins by acrolein and inflammatory mediators. *Am. J. Physiol.* **276**(4 Pt 1): L549–L555.
57. Deshmukh, H.S. *et al.* 2005. Metalloproteinases mediate mucin 5AC expression by epidermal growth factor receptor activation. *Am. J. Respir. Crit. Care Med.* **171:** 305–314.
58. Deshmukh, H.S. *et al.* 2009. Matrix metalloproteinase-14 mediates a phenotypic shift in the airways to increase mucin production. *Am. J. Respir. Crit. Care Med.* **180:** 834–845.
59. Liu, D.S. *et al.* 2009. p38 MAPK and MMP-9 cooperatively regulate mucus overproduction in mice exposed to acrolein fog. *Int. Immunopharmacol.* **9:** 1228–1235.
60. Deshmukh, H.S. *et al.* 2008. Acrolein-activated matrix metalloproteinase 9 contributes to persistent mucin production. *Am. J. Respir. Cell Mol. Biol.* **38:** 446–454.
61. Ren, S. *et al.* 2011. Doxycycline attenuates acrolein-induced mucin production, in part by inhibiting MMP-9. *Eur. J. Pharmacol.* **650:** 418–423.
62. Chen, Y.J. *et al.* 2010. Simvastatin attenuates acrolein-induced mucin production in rats: involvement of the Ras/extracellular signal-regulated kinase pathway. *Int. Immunopharmacol.* **10:** 685–693.
63. Shapiro, S.D. & E.P. Ingenito. 2005. The pathogenesis of chronic obstructive pulmonary disease: advances in the past 100 years. *Am. J. Respir. Cell Mol. Biol.* **32:** 367–372.
64. Demkow, U. & F.J. van Overveld. 2010. Role of elastases in the pathogenesis of chronic obstructive pulmonary disease: implications for treatment. *Eur. J. Med. Res.* **15**(Suppl 2): 27–35.
65. Vernooy, J.H. *et al.* 2004. Increased activity of matrix metalloproteinase-8 and matrix metalloproteinase-9 in induced sputum from patients with COPD. *Chest* **126:** 1802–1810.
66. Molet, S. *et al.* 2005. Increase in macrophage elastase (MMP-12) in lungs from patients with chronic obstructive pulmonary disease. *Inflamm. Res.* **54:** 31–36.
67. Oikonomidi, S. *et al.* 2009. Matrix metalloproteinases in respiratory diseases: from pathogenesis to potential clinical implications. *Curr. Med. Chem.* **16:** 1214–1228.
68. Lemaître, V., A.J. Dabo & J. D'Armiento. 2011. Cigarette smoke components induce matrix metalloproteinase-1 in aortic endothelial cells through inhibition of mTOR signaling. *Toxicol. Sci.* **123:** 542–549.
69. O'Toole, T.E. *et al.* 2009. Acrolein activates matrix metalloproteinases by increasing reactive oxygen species in macrophages. *Toxicol. Appl. Pharmacol.* **236:** 194–201.
70. Kim, C.E. *et al.* 2010. Acrolein increases 5-lipoxygenase expression in murine macrophages through activation of ERK pathway. *Toxicol. Appl. Pharmacol.* **245:** 76–82.
71. Borchers, M.T. *et al.*, 2007. CD8+ T cells contribute to macrophage accumulation and airspace enlargement following repeated irritant exposure. *Exp. Mol. Pathol.* **83:** 301–310.
72. Witschi, A. *et al.* 1992. The systemic availability of oral glutathione. *Eur. J. Clin. Pharmacol.* **43:** 667–669
73. Rahman, I. 2011. Pharmacological antioxidant strategies as therapeutic interventions for COPD. *Biochim. Biophys. Acta.* Nov **9.**
74. Aldini, G. *et al.* 2007. Intervention strategies to inhibit protein carbonylation by lipoxidation-derived reactive carbonyls. *Med. Res. Rev.* **27:** 817–868.
75. Cai, S. *et al.* 2009. Oral N-acetylcysteine attenuates pulmonary emphysema and alveolar septal cell apoptosis in smoking-induced COPD in rats. *Respirology* **14:** 354–359.
76. Stav, D. & M. Raz. 2009. Effect of N-acetylcysteine on air trapping in COPD: a randomized placebo-controlled study. *Chest* Aug; **136:** 381–386.
77. Black, P.N. *et al.* 2004. Randomised, controlled trial of N-acetylcysteine for treatment of acute exacerbations of chronic obstructive pulmonary disease. *BMC Pulm. Med.* **6:** 13.
78. Schermer, T. *et al.* 2009. Fluticasone and N-acetylcysteine in primary care patients with COPD or chronic bronchitis. *Respir. Med.* **103:** 542–551.
79. Decramer, M. *et al.* 2005. Effects of N-acetylcysteine on outcomes in chronic obstructive pulmonary disease (Bronchitis Randomized on NAC Cost-Utility Study, BRONCUS): a randomised placebo-controlled trial. *Lancet* **365:** 1552–1560.
80. Olsson, B. *et al.* 1988. Pharmacokinetics and bioavailability of reduced and oxidized N-acetylcysteine. *Eur. J. Clin. Pharmacol.* **34:** 77–82.
81. Aitio, M.L. 2006. N-acetylcysteine – passe-partout or much ado about nothing? *Br. J. Clin. Pharmacol.* **61:** 5–15.
82. Sadowska, A.M., B. Manuel-Y-Keenoy & W.A. De Backer. 2007. Antioxidant and anti-inflammatory efficacy of NAC

in the treatment of COPD: discordant in vitro and in vivo dose-effects: a review. *Pulm. Pharmacol. Ther.* **20:** 9–22
83. Burcham, P.C. & S.M. Pyke. 2006. Hydralazine inhibits rapid acrolein-induced protein oligomerization: role of aldehyde scavenging and adduct trapping in cross-link blocking and cytoprotection. *Mol. Pharmacol.* **69:** 1056–1065.
84. Wondrak, G.T. *et al.* 2002. Identification of alpha-dicarbonyl scavengers for cellular protection against carbonyl stress. *Biochem. Pharmacol.* **63:** 361–373.
85. Vistoli, G. *et al.* 2009. Design, synthesis, and evaluation of carnosine derivatives as selective and efficient sequestering agents of cytotoxic reactive carbonyl species. *Chem. Med. Chem.* **4:** 967–975.
86. Zhu, Q. *et al.* 2011. Acrolein scavengers: reactivity, mechanism and impact on health. *Mol. Nutr. Food Res.* **55:** 1375–1390.
87. Higgins, L.G. *et al.* 2009. Transcription factor Nrf2 mediates an adaptive response to sulforaphane that protects fibroblasts in vitro against the cytotoxic effects of electrophiles, peroxides and redox-cycling agents. *Toxicol. Appl. Pharmacol.* **237:** 267–280.
88. Kode, A. *et al.* 2008. Resveratrol induces glutathione synthesis by activation of Nrf2 and protects against cigarette smoke-mediated oxidative stress in human lung epithelial cells. *Am. J. Physiol. Lung Cell Mol. Physiol.* **294:** L478–L488.
89. Balstad, T.R. *et al.* 2011. Coffee, broccoli and spices are strong inducers of electrophile response element-dependent transcription in vitro and in vivo—studies in electrophile response element transgenic mice. *Mol. Nutr. Food Res.* **55:** 185–197.
90. Churg, A., D.D. Sin & J.L. Wright. 2011. Everything prevents emphysema: are animal models of cigarette smoke-induced chronic obstructive pulmonary disease any use? *Am. J. Respir. Cell Mol. Biol.* **45:** 1111–1115.
91. Churg, A. & J.L. Wright. 2009. Testing drugs in animal models of cigarette smoke-induced chronic obstructive pulmonary disease. *Proc. Am. Thorac. Soc.* **6:** 550–552.
92. Churg, A., M. Cosio & J.L.Wright. 2008. Mechanisms of cigarette smoke-induced COPD: insights from animal models. *Am. J. Physiol. Lung Cell Mol. Physiol.* **294:** L612–L631.

Ann. N.Y. Acad. Sci. ISSN 0077-8923

ANNALS OF THE NEW YORK ACADEMY OF SCIENCES
Issue: *Environmental Stressors in Biology and Medicine*

Sarcopenia and smoking: a possible cellular model of cigarette smoke effects on muscle protein breakdown

Oren Rom,[1] Sharon Kaisari,[1] Dror Aizenbud,[1,2] and Abraham Z. Reznick[1]

[1]Department of Anatomy and Cell Biology, Rappaport Faculty of Medicine, Technion–Israel Institute of Technology, Haifa, Israel. [2]Orthodontic and Craniofacial Department, Rambam Health Care Campus, Haifa, Israel

Address for correspondence: Abraham Reznick, Department of Anatomy and Cell Biology, Rappaport Faculty of Medicine, Technion, Israel, Efron St., P.O.Box 9649, Bat Galim, Haifa 31096, Israel. Reznick@tx.technion.ac,il

Sarcopenia, the age-related loss of muscle mass and strength, is a multifactorial impaired state of health. Lifestyle habits such as physical activity and nutrition have a major impact on sarcopenia progression. Several epidemiological studies have also shown an association between cigarette smoking and increased levels of sarcopenia in elderly long-time smokers. Clinical, *in vivo*, and *in vitro* studies have tried to investigate the mechanism behind exposure to cigarette smoke (CS) and the subsequent effects on skeletal muscles. The aim of this review is to present a cellular model of CS-induced skeletal muscle protein breakdown based on recent studies dealing with this issue and to propose new potential research directions that may explain the effects of exposure to CS on skeletal muscle integrity.

Keywords: cigarette smoke; sarcopenia; MAPKs; NF-κB; MAFbx/atrogin-1; MuRF1

Introduction

The term *sarcopenia* describes the loss of skeletal muscle mass and strength that occurs with advancing age. Sarcopenia represents an impaired state of health with a high-personal toll—increased risk of falls and fractures, loss of independence, mobility disorders, and increased risk of death.[1] The prevalence of sarcopenia in the United States and parts of Europe has been reported to be 5–13% in people aged 60–70 years and 11–50% in those older than 80 years.[2] With increased aging of the population, the prevalence of sarcopenia is likely to increase even further. Therefore, understanding the risk factors for sarcopenia and strategies for prevention of sarcopenia are of considerable importance for the well-being of the aging population.

Both intrinsic and extrinsic factors contribute to the development of sarcopenia. Intrinsic factors include reduction in anabolic hormones, for example, testosterone, estrogen, growth hormone, and insulin-like growth factor-1 (IGF-1); increased inflammatory activity, as measured by cytokines such as interlukin-6 (IL-6) and tumor necrosis factor-α (TNF-α), which contribute to muscle catabolism; accumulation of free radicals and oxidative stress with increasing age; and changes of mitochondrial function of muscle cells. Extrinsic factors include impaired nutrition, reduced physical activity, catabolic medications, immobility, and illnesses. Some pathological conditions have been also associated with sarcopenia. For example, chronic obstructive pulmonary disease (COPD) patients have been identified with high prevalence of sarcopenia.[3]

Oxidative stress and chronic inflammation in aged skeletal muscle activate signaling pathways that are related to the imbalance between protein synthesis and breakdown, leading to fiber loss and atrophy and eventually to sarcopenia. One of these major pathways is the nuclear factor κB (NF-κB) pathway, which modulates immune response, inflammation, survival, and proliferation. NF-κB regulates myogenic activity and molecules associated with muscle atrophy.[4] Reactive oxygen species (ROS) and TNF-α both activate NF-κB. Inhibition of the NF-κB pathway in atrophic models prevents muscle degeneration and myofiber death.[4] NF-κB protein levels were found to be fourfold higher in elderly skeletal muscles compared to those of young people. This increased concentration is accompanied by anabolic

doi: 10.1111/j.1749-6632.2012.06532.x

signaling deficits observed in wasting muscle.[4] Another relevant pathway in muscle catabolism is the mitogen-activated protein kinases (MAPKs) pathway. p38, extracellular signal-regulated kinase 1/2 (ERK1/2), and c-Jun NH_2-terminal kinase (JNK) are all activated in myotubes exposed to either TNF-α or H_2O_2. Higher levels of ERK1/2 were found in aged skeletal muscles compared to young muscles. Moreover, p38 MAPK signaling has been shown to promote the expression of muscle-specific E3 ubiquitin-ligating enzyme of the ubiquitin proteasome system (UPS).[4] The UPS mediates a large part of the degradation of myofibrillar proteins in skeletal muscle. Ubiquitin addition to a protein substrate requires three components: E1 ubiquitin-activating enzyme, E2 ubiquitin-conjugating enzyme, and E3 ubiquitin-ligating enzyme. E3s play an important role in determining which proteins are targeted for degradation by the proteasome. Two muscle-specific E3s that are overexpressed in numerous catabolic states have been identified: muscle atrophy F-box protein (MAFbx/atrogin-1) and muscle RING finger-1 protein (MuRF1). Mice in which either E3 is knocked out were partially resistant to muscle atrophy.[4] Cbl-b is another RING-type ubiquitin ligase that have been shown to be upregulated in various conditions of skeletal muscle atrophy such as unloading stress and microgravity during spaceflight.[5,6] Cbl-b has been shown to downregulate the IGF-1 signaling pathway in skeletal muscle under unloading conditions. Additionally, Cbl-b–deficient mice were resistant to unloading-induced atrophy and the loss of muscle function.[5]

Tobacco smoke is probably the single most significant source of toxic chemical exposure to humans. Smoking is associated with increased incidence of cardiovascular diseases including coronary heart diseases, cerebrovascular diseases, and vascular diseases, and it is the primary cause of COPD.[7] Tobacco use still remains the single largest preventable cause of disease and premature death in the United States, accounting for at least 30% of cancer deaths.[8]

Mainstream cigarette smoke (CS) is a complex aerosol consisting of a vapor and particulate phase. Some components—for example, carbon monoxide, acrolein, and nitrogen oxides—are found primarily in the vapor phase, while others, such as nicotine, predominate in the particulate phase.[8] Among the thousands of different constituents of mainstream CS are also several types of oxygen- and nitrogen-free radicals that have been associated with numerous pathologies. The gas phase contains more than 10^{15} free radicals per puff, while the particulate phase contains more than 10^{17} of free radicals per gram.[7] Gas phase CS contains ROS—e.g., hydrogen peroxide, superoxide, and hydroxyl radicals—which induce oxidative stress. Additionally, CS is the largest source of nitric oxide (NO). NO and superoxide may react to form peroxynitrite ($ONOO^-$), a potent oxidizing and nitrating compound that also has been linked to a variety of pathological conditions.[9]

In addition to the known harmful effects of cigarette smoking, several studies have identified cigarette smoking as a risk factor for sarcopenia. In the MINOS cohort study, which included 845 men aged 45–85, smokers had lower relative appendicular skeletal muscle mass, as measured by dual-energy X-ray absorptiometry (DEXA), than did subjects who had never smoked.[10] Additionally, the Rancho Bernardo cohort study, which examined sarcopenia prevalence and risk factors in 1,700 men and women aged 55–98, identified tobacco use as a reversible risk factor for sarcopenia.[11] In a similar work undertaken to study the association between sarcopenia and lifestyle factors, sarcopenia was associated with cigarette smoking in 4,000 elderly Chinese.[12] However, the cellular and molecular mechanisms leading to CS-associated muscle breakdown still remains elusive. Nonetheless, recent studies shed some clues and new information that may explain the mechanisms behind CS-induced muscle protein degradation.

The cellular model of CS-induced muscle breakdown

Based on recent studies dealing with the effect of CS on skeletal muscle, we propose a cellular model of CS-induced skeletal muscle protein breakdown (Fig. 1). According to the proposed model, volatile and soluble components of CS, which include aldehydes, ROS, and reactive nitrogen species (RNS), enter the bloodstream and reach the skeletal muscles of smokers. In skeletal muscle, components of CS increase oxidative stress either directly or by activation of nicotinamide adenine dinucleotide phosphate (NADPH) oxidase (NOX) to produce ROS and increased oxidative stress. CS-induced oxidative stress may lead to phosphorylation of p38 MAPK, which in turn activates the NF-κB

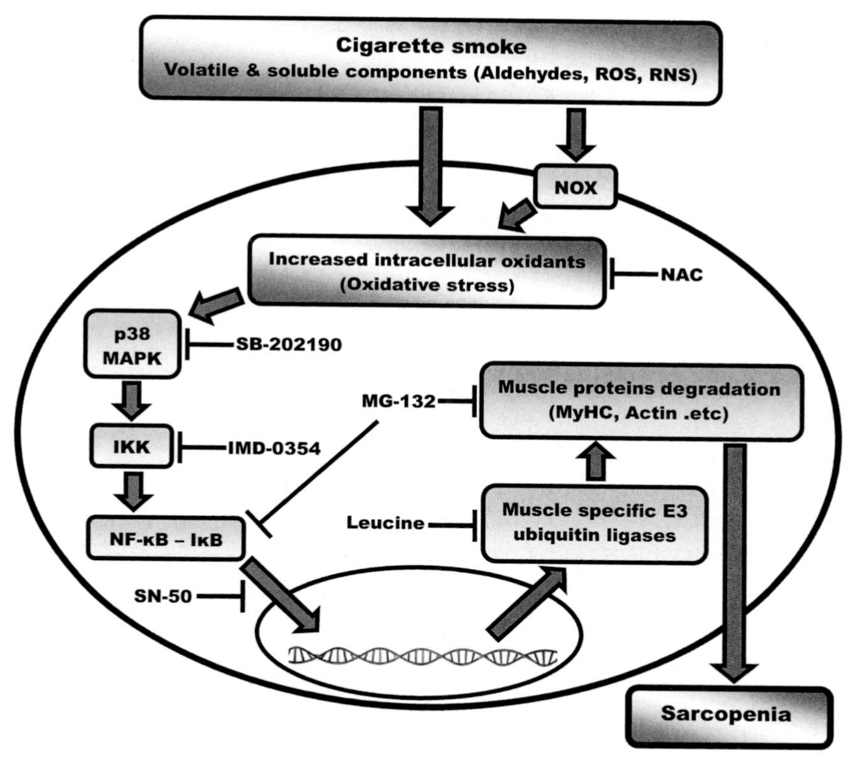

Figure 1. Cellular model of CS-induced muscle breakdown. Nicotinamide adenine dinucleotide phosphate oxidase (NOX); NAC, *N*-acetylcysteine–antioxidant and α,β-unsaturated aldehyde scavenger; p38 MAPK, mitogen-activated protein kinase p38; SB-202190, selective p38 MAPK inhibitor; IKK, IκB kinase; IMD-0354, selective IKK inhibitor; NF-κB, nuclear factor κB; IκB, inhibitor of NF-κB; SN-50, nuclear translocation inhibitor of NF-κB; MG-132, selective proteasome, and IκB degradation inhibitor.

pathway through phosphorylation of inhibitor of NF-κB kinase (IKK) and inhibitor of NF-κB (IκB), and protesomal degradation of IκB, thereby resulting in nuclear translocation of NF-κB. Activated NF-κB triggers upregulation of the muscle-specific E3 ubiquitin ligases. Upregulation of these ligases will lead to increased skeletal muscle protein degradation, thus accelerating the progression of sarcopenia in elderly smokers.

CS-induced oxidative stress in skeletal muscle

Several studies indicate an increase in skeletal muscle oxidative stress parameters following exposure to CS. Barreiro *et al.*[13] have investigated oxidative stress in skeletal muscles of human smokers and guinea pigs chronically exposed to CS. Compared with control subjects, protein oxidation and carbonylated contractile protein levels were increased in quadriceps of smokers and patients with COPD and in limb muscles of CS-exposed animals. Additionally, muscle force of quadriceps was mildly but significantly reduced in smokers compared with control subjects. These results have led to the conclusion that CS causes direct oxidative damage to muscle proteins, which might contribute to muscle loss and dysfunction in smokers and in patients with COPD. Additionally, Ardite *et al.*[14] have found decreased levels of skeletal muscle glutathione (GSH) and increased plasma lipid peroxidation after guinea pigs experienced acute CS exposure. Chronically CS-exposed animals exhibited lower body weight gain. Additionally, Ardite *et al.*[14] concluded that CS exposure induces a transient oxidative effect, which might result in a failure of CS-exposed animals to gain weight.

NOX is a potential inducible source of ROS. In resting cells, NOX has a low level of activity and its protein components are segregated into plasma and cytoplasmic membrane compartments. Once activated, NOX catalyzes the transfer of electrons from NADPH to molecular oxygen with subsequent production of superoxide radical.[15] It was previously shown that aging resulted in increased oxidant production by NOX in rats' skeletal muscle.[16] Previous studies have found that CS extracts (CSE) and water-soluble components of CS are responsible for ROS production via activation of NOX in various cell lines.[15,17] Therefore, it is likely that increased oxidative stress in skeletal muscle following exposure to CS is also mediated by the activation of NOX. Interestingly, Sriram *et al.*[18] have shown that myostatin, a transforming growth factor-β (TGF-β) family member with an important role in skeletal muscle wasting, induces ROS production and oxidative stress in skeletal muscle cells through NOX and NF-κB activation. Additionally, it was shown that myostatin-induced ROS production in C2C12 skeletal muscle cells resulted in upregulation of E3 ubiquitin ligases MAFbx/atrogin-1 and MuRF.[18]

p38 MAPK activation by oxidative stress

The MAPKs signaling pathways are well known to be affected by different stressors placed on the cell. One type of stress that induces phosphorylation and potential kinase activation within the main MAPK pathways is oxidative stress. The p38 MAPK pathway is known to be activated in a number of different cell types in response to reactive oxygen intermediates, such as singlet oxygen, hydrogen peroxide (H_2O_2), nitric oxide, and peroxynitrite.[19] Several studies have investigated the effect of oxidative stress on p38 MAPK activation in skeletal muscle. McClung *et al.*[20] have reported that increased oxidative stress in the skeletal muscle of mice exposed to lipopolysaccharide, and C2C12 myotubes treated with H_2O_2, led to p38 MAPK activation and upregulation of MAFbx/atrogin-1. Inhibition of p38 by SB-202190 blocked hydrogen peroxide–induced muscle atrophy and diminished upregulation of MAFbx/atrogin-1. Similarly, Diamond-Stanic *et al.*[21] have shown that H_2O_2 exposure to isolated rat soleus muscle caused phosphorylation of p38 MAPK.

MAPKs pathway activation following CS exposure

It appears that MAPKs activation is one of the first steps of CS-induced muscle protein breakdown. Liu *et al.*[22] have incubated L6 rat myotubes with CSE and investigated the mechanism underlying the downregulation of myosin heavy chain (MyHC) following CSE incubation. Exposing myotubes to CSE resulted in the general activation of MAPKs, such as p38, JNK, and ERK1/2. In addition to the downregulation of MyHC, CSE exposure caused upregulation of ubiquitin-specific protease-19, which is associated with muscle wasting in several degradative conditions, such as fasting, diabetes, and cancer. Liu *et al.*[22] have concluded that CSE promotes myotube wasting via p38 and ERK MAPKs. Recent findings from our lab support these results.[23] We have investigated the effect of direct exposure of gas phase CS to skeletal muscle C2 myotubes and found CS-induced p38 MAPK phosphorylation followed by degradation of MyHC and actin proteins. It should be noted that several studies have shown MAPKs activation following CS exposure in heart muscle and smooth muscles as well. Gu *et al.*[24] have investigated the effect of CS exposure on MAPKs activation in rat myocardium and found significant increases in phosphorylated p38 and ERK1/2 in hearts of CS-exposed rats. Additionally, exposure of human airway smooth muscle cells to CSE led to a rapid and lasting phosphorylation of p38 MAPK, which was abolished by the antioxidant and α,β-unsaturated aldehyde scavenger *N*-acetylcysteine (NAC),[25] and incubation of human aortic smooth muscle cell cultures with soluble smoke particles caused activation of ERK1/2.[26]

CS-induced NF-κB activation

NF-κB has been shown to be present in the cytoplasm of every cell type in its inactive state. In resting cells, NF-κB, consisting mostly of p50 and p65 subunits, is sequestered in the cytoplasm in an inactive form through its association with its intracellular inhibitor, IκB. In response to environmental stimuli, the inactive NF-κB/IκB complex is activated by phosphorylation by the IKK enzyme on two conserved serine residues within the N-terminal domain of IκB protein. This phosphorylation leads to the immediate poly-ubiquitination of IκB and subsequently targets IκB protein for rapid degradation by the 26S proteasome. Degradation of IκB allows the translocation of NF-κB to the nucleus, where it induces transcription of over 400 genes involved in immune regulation, growth regulation, inflammation, carcinogenesis, and apoptosis. Extensive research has shown that a large number of stimuli can activate NF-κB, including CS.[27] Anto *et al.*[28] have investigated whether CS condensate (CSC) activates NF-κB in a wide variety of cell lines. They have shown that through IKK activation, phosphorylation, and degradation of IκBα, CSC can activate NF-κB in a wide a variety of cells. Similarly, Hasnis *et al.*[9,29,30] have studied the effect of gas phase CS exposure on NF-κB activation in human lymphocytes and found CS-induced NF-κB activation by IKK phosphorylation and IκBα degradation. Recently, we have examined the effect of gas phase CS exposure on NF-κB in C2 skeletal muscle myotubes.[23] In accordance with findings in other cell lines, CS exposure caused NF-κB activation, as seen by IκBα phosphorylation and degradation and by decreased p65 cytoplasmatic levels. Activation of NF-κB was followed by MyHC and actin protein breakdown. The proteasome and NF-κB activation inhibitor MG-132 decreased IκBα phosphorylation and degradation and the subsequent MyHC and actin protein breakdown.

Activation of NF-κB by p38 MAPK

p38 MAPK and NF-κB are known to be activated in response to similar stimuli such as oxidative stress and inflammatory cytokines. A biochemical cross-talk between p38 MAPK and NF-κB has been described in previous studies. Baeza-Raja and Muñoz-Cánoves have shown that p38 MAPK activation induces NF-κB transcriptional activity in C2C12 myoblasts by stimulating IκBα degradation and subsequent nuclear translocation and DNA binding of NF-κB.[31] Additionally, Gomez-Cabrera *et al.*[32] have found that increased levels of ROS and RNS during exercise caused an activation of p38 and ERK 1/2 MAPKs, which in turn activated NF-κB in rat gastrocnemius muscle. Inhibition of ROS and RNS production prevented activation of both MAPKs and the NF-κB pathway. Gochman *et al.*[33] have also shown that NF-κB is activated by p38 MAPK in response to $ONOO^-$. $ONOO^-$ exposure to HT-29 cells caused p38 MAPK activation, which in turn led to NF-κB activation through IKK phosphorylation, IκBα phosphorylation, and subsequent NF-κB nuclear translocation. Inhibition of p38 MAPK activation by p38-specific inhibitor SB-202190 prior to $ONOO^-$ exposure prevented NF-κB activation. Therefore, it appears that p38 MAPK activation following CS exposure to skeletal muscle can lead to NF-κB pathway activation, which promotes muscle protein breakdown.

p38 and NF-κB upregulate muscle-specific E3 ligases

MAFbx/atrogin-1 and MuRF1 are two muscle-specific E3 ligases that are overexpressed in numerous catabolic states. MAFbx/atrogin-1 and MuRF1 target specific muscle proteins for polyubiquitination and subsequent proteasomal degradation during skeletal muscle atrophy. MAFbx/atrogin-1 has been shown to target myogenic transcription factors such as MyoD and myogenin, and MuRF1 preferentially interacts with structural proteins such as MyHC and myosin light chain.[34] MuRF-1 transcription is believed to be driven by the activation of NF-κB.[4] Activation of NF-κB in muscle-specific transgenic expression of activated IKK (MIKK) mice by Cai *et al.*[35] has been demonstrated to induce significant atrophy through expression of MuRF1, but not MAFbx/atrogin-1. They have proposed a signaling pathway of IKK/NF-κB/MuRF1 in which atrophic stimuli activate NF-κB, which in turn results in upregulation of MuRF1. Cai *et al.*[35] have demonstrated that MAFbx/atrogin-1 upregulation was not required for NF-κB–induced muscle loss. Li *et al.*[36] have shown that both H_2O_2 and TNF-α exposure to C2C12 myotubes and to adult mice stimulated MAFbx/atrogin-1 gene expression through the activation of p38 MAPK. MAFbx/atrogin-1 upregulation was blunted by p38 inhibitors SB-203580 and curcumin, suggesting that p38 MAPK is the trigger for upregulation of MAFbx/atrogin-1.[36] As mentioned above, a cross-talk between p38 MAPK and NF-κB has been previously proposed. Indeed, Li *et al.*[37] have shown that the inflammatory cytokine IL-1 stimulated phosphorylation of p38 MAPK and NF-κB activation, resulting in increased expression of both MAFbx/atrogin-1 and MuRF1 and reduced myofibrillar protein in C2C12 myotubes. Thus, it appears that p38 activation induces a cross-effect resulting in direct MAFbx/atrogin-1 upregulation and indirect MuRF1 upregulation through activation of the NF-κB pathway.

CS exposure effect on muscle-specific E3 ligases

Several studies examined the effect of CS or smoking on muscle-specific E3s ligases. Petersen *et al.*[38] have studied the effect of smoking on skeletal muscle protein metabolism. They have found lower muscle protein synthesis rates and higher MAFbx/atrogin-1 expression in the muscle of smokers compared with that of non-smokers and concluded that smoking impairs the muscle protein synthesis processes and increases the expression of genes associated with impaired muscle. Additionally, Tang *et al.*[39] have exposed mice to CS daily for eight or 16 weeks and found increased mRNA levels of both MAFbx/atrogin-1 and MuRF1 following these periods of CS exposure. Additionally, mice exposed to CS exhibited various damages to skeletal muscle, including muscle atrophy. Recently, we have examined the effect of gas phase CS exposure on MAFbx/atrogin-1 and MuRF1 mRNA and protein levels in C2 skeletal muscle myotubes.[23] We have found that CS exposure caused upregulation of both MAFbx/atrogin-1 and MuRF1 that was followed by MyHC and breakdown of actin proteins. Thus, the findings described above support the notion that the damage and protein breakdown in skeletal muscle following exposure to CS is mediated by the upregulation and activation of muscle-specific E3s of the UPS. Future studies should examine the involvement of other atrophy related genes of the UPS, such as Cb1-b, following exposure to CS in skeletal muscle.[5,6]

Essential amino acids (EAAs) are of great importance for muscle protein synthesis. A number of studies suggest that leucine may be the most critical EAA for stimulation of muscle protein synthesis in older adults.[40] Leucine is an insulin secretagogue and potent activator the mammalian target of rapamycin (mTOR), an energy-sensing signaling pathway in skeletal muscle, which is involved in protein synthesis.[3] Leucine supplementation in rats attenuates muscle wasting induced by immobilization via minimizing gene expression of MAFbx/atrogin-1 and MuRF1.[41] Additionally, leucine decreased MAFbx/atrogin-1 and MuRF1 mRNA levels via the mTOR signaling pathway in C2C12 muscle cells.[42] Thus, it is possible that leucine may attenuate CS-induced skeletal muscle breakdown by downregulation of muscle-specific E3s ligases and increased muscle protein synthesis.

Conclusion

The proposed cellular model of the effects of CS exposure on muscle protein breakdown is based on a plethora of results published in the last years in the literature as well as data obtained in our laboratory. These data have been accumulated from *in vitro*, *in vivo*, and clinical human studies that support the notion that gas phase CS contains compounds that are capable of initiating muscle protein breakdown.

In addition, the model also proposes future approaches to validate the scheme by employing specific pharmacological inhibitors that can prove each step of the model by inhibiting the downstream stages of the cascade (Fig. 1).

Acknowledgments

This work was supported by grants from the Rappaport Institute, the Krol Foundation in Barnegat, New Jersey, the Myers-JDC-Brookdale Institute of Gerontology and Human Development, and ESHEL, the association for planning and development of services for the aged in Israel.

Conflicts of interest

The authors declare no conflicts of interest.

References

1. Cruz-Jentoft, A. *et al.* 2010. Sarcopenia: European consensus on definition and diagnosis: report of the European working group on sarcopenia in older people. *Age Ageing.* **39:** 412–423.
2. Waters, D.L. *et al.* 2010. Advantages of dietary, exercise-related, and therapeutic interventions to prevent and treat sarcopenia in adult patients: an update. *Clin. Interv. Aging.* **5:** 259–270.
3. Visvanathan, R. & I. Chapman. 2010. Preventing sarcopenia in older people. *Maturitas.* **66:** 383–388.
4. Meng, S.J. & L.J Yu. 2010. Oxidative stress, molecular inflammation, and sarcopenia. *Int. J. Mol. Sci.* **11:** 1509–1526.
5. Nakao, R. *et al.* 2009. Ubiquitin ligase Cbl-b is a negative regulator for insulin-like growth factor 1 signaling during muscle atrophy caused by unloading. *Mol. Cell. Biol.* **29:** 4798–4811.
6. Nikawa, T. *et al.* 2004. Skeletal muscle gene expression in space-flown rats. *Faseb. J.* **18:** 522–524.
7. Swan, E.G. & C.N. Lessov-Schlaggar. 2007. The effects of tobacco smoke and nicotine on cognition and the brain. *Neuropsychol. Rev.* **17:** 259–273.
8. Smith, C.J. & T.H Fischer. 2011. Particulate and vapor-phase constituents of cigarette mainstream smoke and risk of myocardial infarction. *Atherosclerosis.* **158:** 257–267.
9. Hasnis, E. *et al.* 2007. Mechanisms underlying cigarette smoke-induced NF-kB activation in human lymphocytes: the role of reactive nitrogen species. *J. Physiol. Pharmacol.* **58:** 275–287.
10. Szulc, P. *et al.* 2004. Hormonal and lifestyle determinants of appendicular skeletal muscle mass in men: the MINOS study. *Am. J. Clin. Nutr.* **80:** 496–503.
11. Castillo, E. *et al.* 2003. Sarcopenia in elderly men and women: the Rancho Bernardo study. *Am. J. Prev. Med.* **25:** 226–231.

12. Lee, J.S. *et al.* 2007. Associated factors and health impact of sarcopenia in older Chinese men and women: a cross-sectional study. *Gerontology*. **53:** 404–410.
13. Barreiro, E. *et al.* 2010. Cigarette smoke-induced oxidative stress. A role in chronic obstructive pulmonary disease skeletal muscle dysfunction. *Am. J. Respir. Crit. Care Med.* **182:** 477–488.
14. Ardite, E. *et al.* 2006. Systemic effects of cigarette smoke exposure in the guinea pig. *Respir. Med.* **100:** 1186–1194.
15. Jaimes, E.A. *et al.* 2004. Stable compounds of cigarette smoke induce endothelial superoxide anion production via NADPH oxidase activation. *Arterioscler. Thromb. Vasc. Biol.* **24:** 1031–1036.
16. Bejma, J. & L.L. Ji. 1999. Aging and acute exercise enhance free-radical generation in rat skeletal muscle. *J. Appl. Physiol.* **87:** 465–470.
17. Orosz, Z. *et al.* 2007. Cigarette smoke-induced proinflammatory alterations in the endothelial phenotype: role of NAD(P)H oxidase activation. *Am. J. Physiol. Heart Circ. Physiol.* **292:** H130–H139.
18. Sriram, S. *et al.* 2011. Modulation of reactive oxygen species in skeletal muscle by myostatin is mediated through NF-κB. *Aging Cell.* **10:** 931–948.
19. McCubrey, J.A. *et al.* 2006. Reactive oxygen species-induced activation of the MAP kinase signaling pathways. *Antioxid. Redox. Signal.* **8:** 1775–1789.
20. McClung, J.M. *et al.* 2010. p38 MAPK links oxidative stress to autophagy-related gene expression in cachectic muscle wasting. *Am. J. Physiol. Cell. Physiol.* **298:** 542–549.
21. Diamond-Stanic, M.K. *et al.* 2011. Critical role of the transient activation of p38 MAPK in the etiology of skeletal muscle insulin resistance induced by low-level *in vitro* oxidant stress. *Biochem. Biophys. Res. Commun.* **405:** 439–444.
22. Liu, Q. *et al.* 2011. Cigarette smoke-induced skeletal muscle atrophy is associated with upregulation of USP-19 via p38 and ERK MAPKs. *J. Cell. Biochem.* **112:** 2307–2316.
23. Rom, O. *et al.* 2011. Sarcopenia and smoking: effects of cigarette smoke on muscle proteins breakdown. In *Second International Conference on Environmental Stressors in Biology and Medicine.* Siena, Italy.
24. Gu, L. *et al.* 2008. Cigarette smoke-induced left ventricular remodeling is associated with activation of mitogen-activated protein kinases. *Eur. J. Heart Fai.* **10:** 1057–1064.
25. Volpi, G. *et al.* 2011. Cigarette smoke and α,β-unsaturated aldehydes elicit VEGF release through the p38 MAPK pathway in human airway smooth muscle cells and lung fibroblasts. *Br. J. Pharmacol.* **163:** 649–661.
26. Chen, Q. *et al.* 2010. Cigarette smoke extract promotes human vascular smooth muscle cell proliferation and survival through ERK1/2- and NF-κB–dependent pathways. *The Scientific World Journal* **10:** 2139–2156.
27. Ahv, K.S & B.B. Aggarwal. 2005. Transcription factor NF-κB. A sensor for smoke and stress signals. *Ann. N.Y. Acad. Sci.* **1056:** 218–233.
28. Anto, R.J. *et al.* 2002. Cigarette smoke condensate activates nuclear transcription factor-kappaB through phosphorylation and degradation of IkappaB(alpha): correlation with induction of cyclooxygenase-2. *Carcinogenesis* **23:** 1511–1518.
29. Hasnis, E. *et al.* 2007. Cigarette smoke-induced NF-kappaB activation in human lymphocytes: the effect of low and high exposure to gas phase of cigarette smoke. *J. Physiol. Pharmacol.* **58:** 263–274.
30. Hasnis, E. *et al.* 2006. The role of reactive nitrogen species and cigarette smoke in activation of transcription factor NF-kappaB and implication to inflammatory processes. *J Physiol. Pharmacol.* **57:** 39–44.
31. Baeza-Raja, B & P. Muñoz-Cánoves. 2004. p38 MAPK-induced nuclear factor-kappaB activity is required for skeletal muscle differentiation: role of interleukin-6. *Mol. Biol. Cell.* **15:** 2013–2026.
32. Gomez-Cabrera, M.C. *et al.* 2005. Decreasing xanthine oxidase-mediated oxidative stress prevents useful cellular adaptations to exercise in rats. *J. Physiol.* **15:** 113–120.
33. Gochman, E. *et al.* 2011. NF-κB activation by peroxynitrite through IκBα-dependent phosphorylation versus nitration in colon cancer cells. *Anticancer Res.* **31:** 1607–1617.
34. Foletta, V.C. *et al.* 2011. The role and regulation of MAFbx/atrogin-1 and MuRF1 in skeletal muscle atrophy. *Pflugers Arch.* **461:** 325–335.
35. Cai, D. *et al.* 2004. IKKbeta/NF-kappaB activation causes severe muscle wasting in mice. *Cell* **119:** 285–298.
36. Li, Y.P. *et al.* 2005. TNF-alpha acts via p38 MAPK to stimulate expression of the ubiquitin ligase atrogin1/MAFbx in skeletal muscle. *FASEB J.* **19:** 362–370.
37. Li, W. *et al.* 2009. Interleukin-1 stimulates catabolism in C2C12 myotubes. *Am. J. Physiol. Cell Physiol.* **297:** 706–714.
38. Petersen, A. *et al.* 2007. Smoking impairs muscle protein synthesis and increases the expression of myostatin and MAFbx in muscle. *Am. J. Physiol. Endocrinol. Metab.* **293:** 843–848.
39. Tang, K. *et al.* 2010. TNF-alpha -mediated reduction in PGC-1a may impair skeletal muscle function after cigarette smoke exposure. *J. Cell. Physiol.* **222:** 320–327.
40. Buford, T.W. *et al.* 2010. Models of accelerated sarcopenia: critical pieces for solving the puzzle of age-related muscle atrophy. *Ageing Res. Rev.* **9:** 369–383.
41. Baptista, I.L. *et al.* 2009. Leucine attenuates skeletal muscle wasting via inhibition of ubiquitin ligases. *Muscle Nerve.* **41:** 800–808.
42. Herningtyas, E.H. *et al.* 2008. Branched-chain amino acids and arginine suppress MaFbx/atrogin-1 mRNA expression via mTOR pathway in C2C12 cell line. *Biochim. Biophys. Acta.* **1780:** 1115–1120.

Ann. N.Y. Acad. Sci. ISSN 0077-8923

ANNALS OF THE NEW YORK ACADEMY OF SCIENCES
Issue: *Environmental Stressors in Biology and Medicine*

The link between altered cholesterol metabolism and Alzheimer's disease

Paola Gamba, Gabriella Testa, Barbara Sottero, Simona Gargiulo, Giuseppe Poli, and Gabriella Leonarduzzi

Department of Clinical and Biological Sciences, Faculty of Medicine San Luigi Gonzaga, University of Turin, Turin, Italy

Address for correspondence: Giuseppe Poli, M.D., Department of Clinical and Biological Sciences, University of Torino, San Luigi Gonzaga Hospital, Regione Gonzole 10, 10043 Orbassano, Torino, Italy. giuseppe.poli@unito.it

Alzheimer's disease (AD), the most common form of dementia, is characterized by the progressive loss of neurons and synapses, and by extracellular deposits of amyloid-β (Aβ) as senile plaques, Aβ deposits in the cerebral blood vessels, and intracellular inclusions of hyperphosphorylated tau in the form of neurofibrillary tangles. Several mechanisms contribute to AD development and progression, and increasing epidemiological and molecular evidence suggests a key role of cholesterol in its initiation and progression. Altered cholesterol metabolism and hypercholesterolemia appear to play fundamental roles in amyloid plaque formation and tau hyperphosphorylation. Over the last decade, growing evidence supports the idea that cholesterol oxidation products, known as oxysterols, may be the missing link between altered brain cholesterol metabolism and AD pathogenesis, as their involvement in neurotoxicity, mainly by interacting with Aβ peptides, is reported.

Keywords Alzheimer's disease; cholesterol; oxysterols; amyloid-β; oxidative stress; neurotoxicity

Introduction

Alzheimer's disease (AD), a neurodegenerative disorder, is the most common form of dementia in developed countries. It is a complex and genetically heterogeneous disease, characterized by progressive memory deficit, cognitive impairment, and personality changes, accompanied by specific structural abnormalities in the brain. The main histological features of AD are extracellular deposits of amyloid-β (Aβ) in the form of senile plaques, Aβ deposits in the cerebral blood vessels, and intracellular inclusions of hyperphosphorylated tau in the form of neurofibrillary tangles (NFT). The loss of neurons and synapses in the neocortex, hippocampus and other subcortical regions of the brain is also a common feature of AD.[1,2]

AD begins with the abnormal processing of amyloid precursor protein (APP) by the sequential enzymatic actions of two enzymes of the amyloidogenic pathway: beta-site amyloid precursor protein-cleaving enzyme 1 (BACE1), a β-secretase, and γ-secretase; these actions lead to excess of production and/or reduced clearance of Aβ peptides, which comprise 39–43 amino acids. An imbalance between production and clearance of Aβ in the brain, and their aggregation, causes Aβ to accumulate, and this excess may be the initiating factor in AD. Monomers of $A\beta_{40}$ are usually much more prevalent than the aggregation-prone and damaging $A\beta_{42}$ species, but an increased proportion of $A\beta_{42}$ appears sufficient to cause early onset of AD. Additionally, insoluble oligomers and intermediate amyloids are the most neurotoxic forms of $A\beta_{42}$.[3]

Several mechanisms (e.g., perturbation of brain metabolism, oxidative stress, inflammation, presence of the apolipoprotein E [ApoE] ε4 allele, impaired cholesterol metabolism) contribute to the development and progression of AD. Among these, a growing body of epidemiological and molecular evidence suggests a mechanistic link between cholesterol and AD progression. A number of genes involved in cholesterol homeostasis have been identified as susceptibility loci for sporadic or late-onset

doi: 10.1111/j.1749-6632.2012.06513.x

AD,[4–6] and altered cholesterol metabolism seems to play a fundamental role in the formation of amyloid plaques and in tau hyperphosphorylation.[7,8] In addition, hypercholesterolemia is unanimously recognized to be a risk factor for sporadic AD, a form that accounts for the great majority of cases.[4,9–11] Finally, this evidence is supported by epidemiological studies indicating that cholesterol-lowering agents belonging to the family of statins reduce the prevalence of AD,[12–14] a conclusion not yet fully accepted because of the contradictory results reported by prospective clinical studies.[15–17]

Apolipoprotein E and its receptors in AD

Apolipoprotein E (ApoE) is the brain's principal cholesterol-carrier protein, mainly transporting it from astrocytes to neurons. The association between ApoE polymorphism and AD is presumably related to the disturbance of cholesterol transport. Of note, subjects who are homozygous for the ApoE ε4 genotype express an increased AD risk, versus those carrying ε2 or ε3, evidence that is consistently confirmed by numerous independent studies.[18–21] In addition, receptors recognizing ApoE are also widely expressed in the AD brain.[4,20,22]

Although ApoE mediates Aβ clearance by binding Aβ and forming a stable complex, ApoE may also stimulate Aβ aggregation and amyloid deposition, as well as tau hypersphorylation.[20,22–25] Moreover, among the mechanisms that might explain the effects of ApoE on the brain of AD subjects, ApoE ε4 and its receptors are also reported to be involved in APP trafficking and its processing to Aβ. Additionally, ApoE may mediate Aβ cell internalization, by binding to the LDL receptor–related protein (LRP).[26] ApoE might also modulate the distribution and metabolism of cholesterol in neuronal membranes, and regulate the role of cholesterol in synapse formation and function, through ApoE receptors.[20,25,27] Moreover, γ-secretase cleavage of APP could thus regulate ApoE metabolism through the LRP1 receptor.[27]

A number of epidemiological studies also report that individuals with high levels of blood cholesterol have an increased susceptibility to AD, apparently influenced by the ApoE ε4 genotype, which may influence cholesterol metabolism and the formation of cholesterol oxidation products, known as oxysterols.[28]

Role of cholesterol in AD

The brain is the organ with the highest concentration of cholesterol, which is essential for its normal function, being a major component of neuronal cell membranes and a determinant of membrane fluidity.[29] In the brain, cholesterol is mostly present in the free form and is derived from *de novo* biosynthesis from acetyl-coenzyme A mediated by 3-hydroxy-3-methyl-glutaryl-coenzyme A (HMG-CoA) reductase, rather than from plasma lipoproteins, which are prevented by the blood–brain barrier (BBB) from crossing from the peripheral circulation into the brain. Further, the astrocytic compartment meets neuronal cholesterol demands by secreting ApoE-cholesterol complexes, which are transported to the neurons.[30,31]

The mechanism by which cholesterol affects Aβ production and metabolism is not fully understood; however, a change in membrane properties has been suggested.[9] Cholesterol is mainly concentrated in membrane microdomains termed *lipid rafts* where considerable evidence indicates that the amyloidogenic pathway takes place.[32,33] In this connection, it has been reported that cellular cholesterol, especially when it is elevated in the membrane, binds directly to APP and thus promotes APP's insertion into the phospholipid monolayers of the lipid rafts and other organelles where β- and γ-secretases reside, and favors the amyloidogenic pathway.[34–36] Amyloidogenic activity is thus linked to cholesterol levels: β- and γ-secretase activities are positively regulated by high and inhibited by low levels of cholesterol.[37,38] Since APP processing and Aβ generation are associated with cholesterol-rich microdomains, and both are present in lipid rafts, Aβ production may require raft integrity and a lipid component as optimal conditions. An alteration in raft components could thus change the configuration of either the enzymes or the substrate associated with the rafts, leading to an alteration in Aβ generation. Conversely, in the nonamyloidogenic pathway, APP is processed by α-secretase in nonraft domains, and this event is promoted by a decreased cellular cholesterol level.[39] Moreover, the amyloidogenic pathway is inactivated when α-secretase is forced to associate with lipid rafts.[40]

Cholesterol also enhances Aβ to form neurotoxic aggregates.[41] In addition, fibrillogenesis of Aβ has been proposed to take place in lipid rafts at ganglioside clusters, where Aβ displays a specific affinity to cholesterol, which binds avidly to Aβ protofibrils.[42,43] Moreover, it has been shown that increased cholesterol levels in the lipid bilayers facilitate binding of Aβ to the membranes, promoting Aβ conformational change from a helix-rich to a β-sheet–rich structure, and thus becoming an endogenous seed for amyloid formation.[44] The conversion of soluble and nontoxic monomeric Aβ to insoluble and toxic oligomeric and aggregate Aβ is thus the critical step in AD development.

Given the above considerations, it seems that cholesterol distribution and trafficking within brain cells, rather than total cholesterol levels in the neurons, are the relevant factors in the APP processing and Aβ accumulation in the AD progression.[45]

It has also been observed that high concentrations of free cholesterol alone do not affect APP processing or Aβ production; rather, the conversion of excess free cholesterol into cholesterol ester has a profound effect on APP and Aβ enhancing their upregulation.[46] Consequently, clearance of Aβ from the brain is reduced when an overabundance of esterified cholesterol decreases membrane lipid turnover. Conversely, inhibition of the enzyme acyl-coenzyme A:cholesterol acyl-transferase 1 (ACAT1), which esterifies cholesterol, leads to the reduction of both cholesteryl esters and Aβ.[47] These data suggest that the balance between free cholesterol and cholesterol esters is a key parameter controlling amyloidogenesis, although the molecular mechanisms underlying this relationship are still unclear.

A regulatory role for APP and γ-secretase in cholesterol metabolism, jointly acting to lower cellular cholesterol levels, has also been reported.[27] The effect of Aβ and of the intracellular domain of APP (AICD) on cellular cholesterol metabolism have also been investigated. Aβ, in particular the oligomeric rather than the monomeric form, alters intracellular trafficking and cholesterol homeostasis, by promoting the release of cholesterol and some other lipids from cells, in the form of Aβ-lipid particles.[48] The fibrillar Aβ then downregulates cholesterol biosynthesis.[49] The AICD, which is released upon γ-secretase cleavage of APP, downregulates cellular cholesterol uptake by acting as a transcriptional suppressor of LRP1 gene, a major ApoE receptor in the brain.[27] Furthermore, the peptide products of the amyloidogenic pathway ultimately reduce both cholesterol uptake and its biosynthesis, completing a negative feedback loop. A decrease in cellular cholesterol levels then results in an enhancement of tau phosphorylation[50] and synaptic failure.[51] It has also been demonstrated that extracellular cholesterol accumulates in the senile plaques and neurofibrillary tangles of AD patients, and in transgenic mice expressing the Swedish Alzheimer mutation APP751, as well as ApoE, and that cholesterol, ApoE, and Aβ all colocalize in the core of fibrillar plaques.[52,53] This shows that both ApoE and cholesterol may be essential for plaque formation, and that extracellular cholesterol, by binding to aggregated Aβ, may be the seed for its deposition.

Role of oxysterols in AD

Because there is little synthesis of cholesterol in the adult brain, and because the brain cannot degrade cholesterol, it must be excreted from the brain in order to prevent its accumulation. The most important mechanism whereby the brain eliminates excess cholesterol is through the formation and excretion into the circulation of oxysterols, a class of cholesterol oxidation products, which are thus important to balance the local synthesis of sterols.[54]

Cholesterol is primarily converted into the oxysterol 24-hydroxycholesterol (24-OH), also known as cerebrosterol, which is produced almost exclusively in the brain by CYP46A1 (cholesterol 24-hydroxylase) and which, unlike cholesterol itself, can easily cross the BBB.[9,30,54,55] Following its secretion from the brain, 24-OH enters the circulation and reaches the liver, where it is taken up and metabolized. The oxysterol 24-OH also plays an important role in the regulation of cholesterol homeostasis in the brain. Neuronal cells have a lower rate of cholesterol synthesis than glial cells; for this reason, an increased flux of 24-OH from neurons to glial cells causes an increased flux of cholesterol to the neuronal cells, by means of activation of the nuclear liver X receptor (LXR) and upregulation of ApoE in the glial cells (Fig. 1).[56] Another oxysterol, 27-hydroxycholesterol (27-OH), has been found to be produced *in situ* in the brain by CYP27A1, although in small amounts, and then metabolized by the enzyme CYP7B to 7α-hydroxy-3-oxo-4-cholestenoic acid (7-OH-4-C), which, crossing the BBB, reaches the liver where it is eliminated.[54,55,57] However, it has been observed that most 27-OH flows from the circulation into the brain, since, unlike cholesterol, it can cross the BBB.[58] Summarizing, these

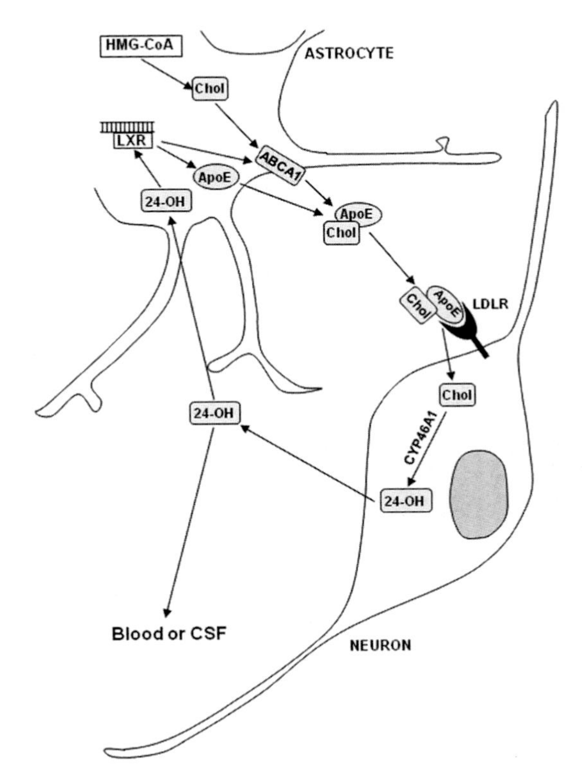

Figure 1. Cholesterol homeostasis in the mature brain. Excess cholesterol is converted into 24-hydroxycholesterol (24-OH) in neuronal cells by CYP46A1. Most of the 24-OH goes directly from the brain into the blood circulation, a small quantity entering the cerebrospinal fluid (CSF). 24-OH may also be caught by astrocytes, where it upregulates the nuclear receptor liver X receptor (LXR)-responsive genes involved in cholesterol efflux: ATP-binding cassette transporter A1 and G1 (ABCA1 and ABCG1) and apolipoprotein E (ApoE). Synthesized cholesterol is loaded by astrocytes onto ApoE, and the ApoE/cholesterol complex is then internalized by neurons via low-density lipoprotein receptors (LDLR).

considerations indicate that there are fluxes of oxysterols in opposite directions across the BBB: two fluxes out of the brain (24-OH and 7-OH-4-C) and one flux into the brain (27-OH). A further compound, 7β-hydroxycholesterol (7β-OH), may also derive in the brain from oxidation of cholesterol following cholesterol interaction with Aβ and APP (Fig. 2).[59]

During the last decade, the idea that oxysterols might be the missing link between altered brain cholesterol metabolism and AD pathogenesis has been increasingly supported by research pointing to the involvement of 24-OH and 27-OH in neurotoxicity, mainly by interacting with Aβ peptides.

Several studies have found higher levels of 24-OH in the peripheral circulation and cerebrospinal fluid (CSF) of AD patients during the early stages than in unaffected individuals, suggesting that cholesterol turnover in the brain increases during the neurodegenerative changes of AD.[60–62] Conversely, plasma levels of 24-OH were decreased in patients with later stages of AD than in the respective controls, suggesting that the rate of cholesterol transport lowers as the disease progresses.[62,63] These contradictory results might be rationalized by considering that increased plasma levels of 24-OH reflect ongoing neurodegeneration and/or demyelinization, whereas decreased plasma levels reflect a selective loss of neurons expressing CYP46A1.[30] However, in glial cells of AD brains there is some ectopic induction of CYP46A1, and consequently some 24-OH production, which may overlap with decreased neuronal expression in the presence of increased glial expression,[64,65] although this induction of CYP46A1 cannot compensate for the loss of 24-hydroxylase activity due to the neuronal degeneration. Another study, however, has found that plasma levels of 24-OH in AD patients are not significantly different than in control subjects.[66]

It has also been reported that, in all brain areas of deceased AD patients, as well as in aged mice expressing the Swedish Alzheimer mutation APP751, the amount of 24-OH decreases and 27-OH increases.[67] A marked accumulation of 27-OH was also found in the brain of patients carrying the Swedish APP670/671 mutation.[68] Whereas the decreased levels of 24-OH in the AD brain are presumably due to the loss of neuronal cells and consequent loss of the enzyme CYP46A1, the increased levels of 27-OH may be due to increased flux of this oxysterol across the BBB, because of hypercholesterolemia,[69] or a damage of BBB integrity.[70] An alternative explanation for the high levels of 27-OH is a reduced metabolism of the oxysterol into 7-OH-4-C, by the enzyme CYP7B, which is reduced in the brain of AD patients.[71] However, increased levels of both 24-OH and 27-OH have been observed in the CSF in patients with advanced AD.[72]

From these considerations, the hypothesis has been formulated that the balance between 24-OH and 27-OH is important for amyloidogenesis,[54,69] and the increased ratio of 27-OH to 24-OH in AD brains is consistent with this hypothesis.[67] Thus, the shift in balance between the two oxysterols might

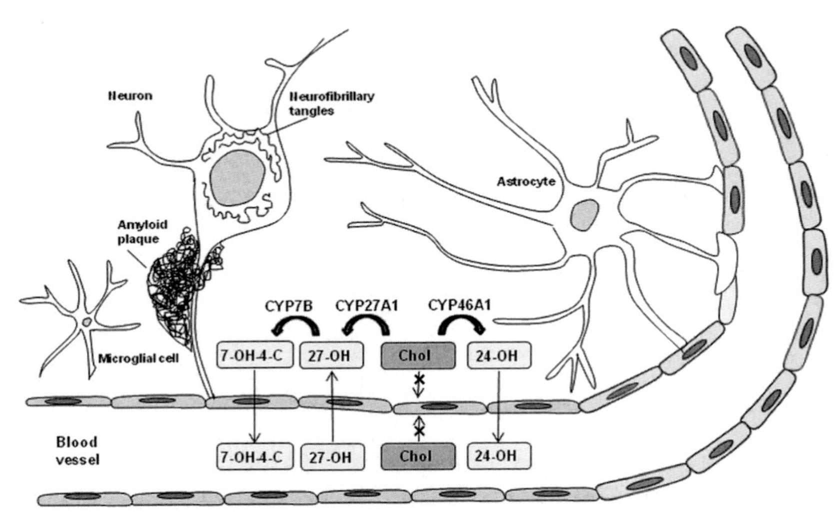

Figure 2. Fluxes of 24-hydroxycholesterol (24-OH) and 27-hydroxycholesterol (27-OH) through the BBB. The enzymes CYP46A1 and CYP27A1, located in the neuronal cells, are respectively responsible for generating 24-OH and 27-OH in the brain. However, most of the 27-OH flows from the circulation into the brain since, unlike cholesterol, it can cross the BBB, as can 24-OH. In the brain, 27-OH is also metabolized to 7α-hydroxy-3-oxo-4-cholestenoic acid (7-OH-4-C), which crosses the BBB to reach the liver. In conclusion, there is a complex of fluxes of oxysterols in opposite directions at the BBB: two fluxes out of the brain (24-OH and 7-OH-4-C) and one flux into the brain (27-OH). Of note, the shift in balance between 24-OH and 27-OH is important for amyloidogenesis since it might lead to increased generation and accumulation of Aβ in AD brain.

lead to increased generation and accumulation of Aβ and regulation of the 24-OH/27-OH ratio could be an important strategy in controlling Aβ levels in AD. However, opinions still differ about the involvement of these two oxysterols in the APP processing and Aβ generation.

According to several studies, induction of CYP46A1 activity has beneficial effects, directly preventing Aβ generation by modulating cholesterol homeostasis and reducing cellular cholesterol. Indeed, astrocytes are sensitive to 24-OH-mediated upregulation of the LXR-responsive genes involved in cholesterol efflux: ATP-binding cassette transporter A1 and G1 (ABCA1 and ABCG1) and ApoE.[56] Conversely, the low levels of 27-OH in the brain might also be expected not to affect amyloidogenesis. However, since the flux of 27-OH across the BBB increases under conditions of hypercholesterolemia,[69] or in the case of reduced BBB integrity,[70] the inhibitory effect of 24-OH on Aβ generation is consequently reduced. In this connection, the high flux of 27-OH from the peripheral circulation to the brain, and changes in the brain cholesterol/oxysterol balance, may partially explain the link between hypercholesterolemia and AD.[69] Nevertheless, in murine primary neuronal cells, both 24-OH and 27-OH were found to inhibit Aβ formation and secretion, 24-OH being about 1,000-fold more potent than 27-OH.[65] The study showed that the distribution of the enzymes CYP46A1 and CYP27A1 is altered in the brain of subjects with AD. CYP46A1, whose expression increases in astrocytes and decreases in neurons, is selectively expressed in degenerating neuritis around senile plaques, whereas CYP27A1 expression, which is mostly expressed in neurons but also to a lesser extent in astrocytes and oligodendrocytes, is increased in white-matter oligodendrocytes.[65] Moreover, it has been shown that 27-OH significantly reduces Aβ peptide generation from primary human neurons, not by affecting α-, β-, or γ-secretase, but by upregulating LXR responsive genes (ABCA1, ABCG1, and ApoE).[73] Of note the LXR-mediated gene regulation of cholesterol efflux and metabolism not only modulates neurodegeneration and Aβ peptide transport and clearance but also inflammation in the brain. On the basis of these data, it cannot be excluded that 27-OH, as an LXR ligand, might exert antiamyloidogenic effects by reducing extracellular Aβ and inflammation.[74–77]

By contrast, other studies are consistent with the possibility that 27-OH may accelerate neurodegeneration. In human SH-SY5Y neuroblastoma cells, 24-OH directly increases α-secretase activity as well as elevating the α/β activity ratio, whereas 27-OH counteracts the inhibitory effect of 24-OH on the generation of amyloid.[78] In addition, recent studies underlined the different effects of 24-OH and 27-OH on APP levels and processing in human neuroblastoma SH-SY5Y cells and in the brain tissue: 24-OH may favor the nonamyloidogenic pathway, whereas 27-OH is thought to enhance production of $A\beta_{42}$ by upregulating APP and BACE1, and tau hyperphosphorylation.[79,80] Another study has found that 27-OH increases Aβ accumulation by reducing insulin-like growth factor 1 (IGF1) levels, a neurotrophic factor that promotes neurogenesis and has a neuroprotective effect, in hippocampal slices from adult rabbits;[81] in addition, 27-OH reduces the production of the memory protein activity-regulated cytoskeleton-associated protein (Arc) in mouse brain.[82]

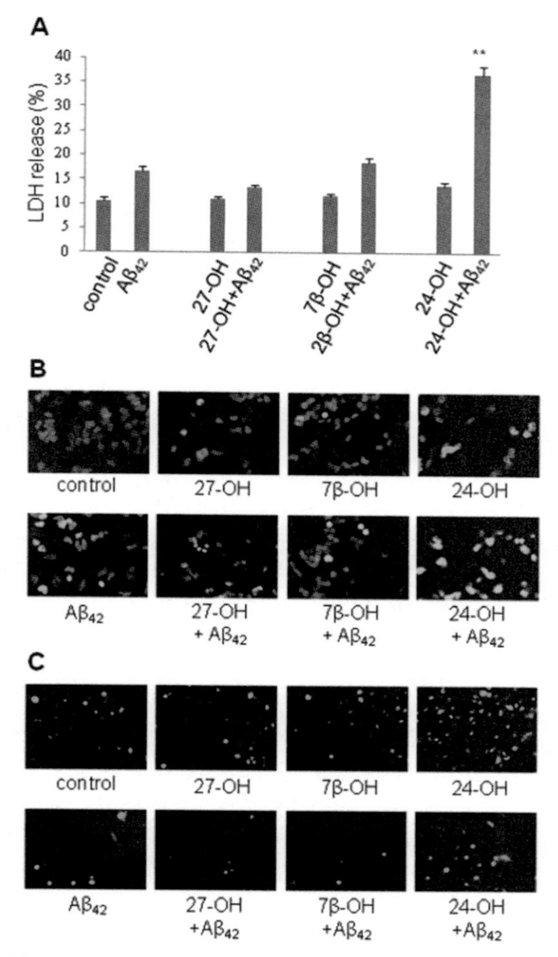

Figure 3. (A) Necrotic effects of 27-hydroxycholesterol (27-OH), 7β-hydroxycholesterol (7β-OH), or 24-hydroxycholesterol (24-OH) evaluated in terms of lactate dehydrogenase (LDH) release. SK-N-BE cells were treated with the oxysterol (1 μM) for 48 h, and then for 24 h with $A\beta_{42}$ (1 μM). Histograms represent the mean values ± SD of three experiments. $^{**}P < 0.01$ versus control (untreated cells). (B) Apoptotic effects of 27-OH, 7β-OH, or 24-OH on SK-N-BE cells observed by DAPI staining to determine apoptotic nuclei formation. Cells were treated with the oxysterol for 48 h and then for 24 h with $A\beta_{42}$. (C) Prooxidant effects of 27-OH, 7β-OH, or 24-OH. Intracellular generation of reactive oxygen species (ROS) was examined in SK-N-BE cells using 2′,7′-dichlorodihydrofluorescein (DCFH-DA) as an intracellular probe. Cells were incubated with oxysterol for 1 h, or simultaneously with oxysterol plus $A\beta_{42}$ for 1 hour.

Alongside altered cholesterol metabolism and hypercholesterolemia, inflammatory response and oxidative stress also significantly contribute to neuronal damage in AD.[83,84]

The importance of inflammatory processes has been pointed out during the past decade by the intensive investigation of inflammatory mediators and microglia activation in the brain of AD, although it remains unclear whether inflammation represents a cause or a consequence of AD. It has been reported that intraneuronal Aβ and soluble Aβ oligomers activate microglia in the earliest stages of the disease, even before plaque and tangle formation, in particular when cells are stressed.[85–87] Fibrillar Aβ can also activate microglia by binding to cells via specific receptors, in particular through a multireceptor complex involving CD36, $\alpha_6\beta_1$-integrin, and CD47.[88] The induction of a microglia-driven inflammatory response results in the release of various inflammatory mediators, including a whole array of neurotoxic cytokines and free radicals.[89] Once activated, microglia cells may also recruit astrocytes, which actively enhance the inflammatory response to extracellular Aβ deposits that intensify neuronal dysfunction and cell death: inflammatory mediators and other components of the immune system are often found near areas of amyloid plaques.[90]

It has also been postulated that oxidative stress may be either a cause or a consequence of the neuropathology associated with AD[91–94] and, in support of its being a consequence, Aβ can stimulate the production of reactive oxygen species (ROS).[95] Conversely, oxidative stress has been shown to contribute to the formation of amyloid plaques,[96] since

Aβ peptide has been found in the oxidized form.[97] Since the brain has a high lipid content, it is extremely vulnerable to free radicals and ROS, which are responsible for enhancing lipid peroxidation, including cholesterol oxidation and oxysterol formation[98] as well as tau hyperphosphorylation and NFT formation,[99] mitochondrial insufficiency, and neuronal cell death.[100]

Oxysterols have been shown to enhance Aβ aggregation and its neurotoxicity, by modifying specific sites of Aβ peptide. Following Aβ modification at Lys-16, peptide aggregates were formed faster than in the case of modification at Lys-28 or Asp-1.[101] Moreover, 7β-OH has been found to be neurotoxic at nanomolar concentrations in cultured rat hippocampal neuronal cells, and may therefore contribute to Aβ-related neurodegeneration in the brain of AD patients.[59] Another oxysterol that might derive from the autooxidation of cellular cholesterol released during neurodegeneration, is 7α-hydroperoxycholesterol, which has also been found to be responsible for necrotic cell death of SH-SY5Y cells,[102] and a further possibility is 7-ketocholesterol.[103] Additionally, 24-OH has been shown to enhance the neurotoxic effect of the $A\beta_{42}$ peptide in the human differentiated neuroblastoma cell line MSN, as well as augmenting ROS generation.[104]

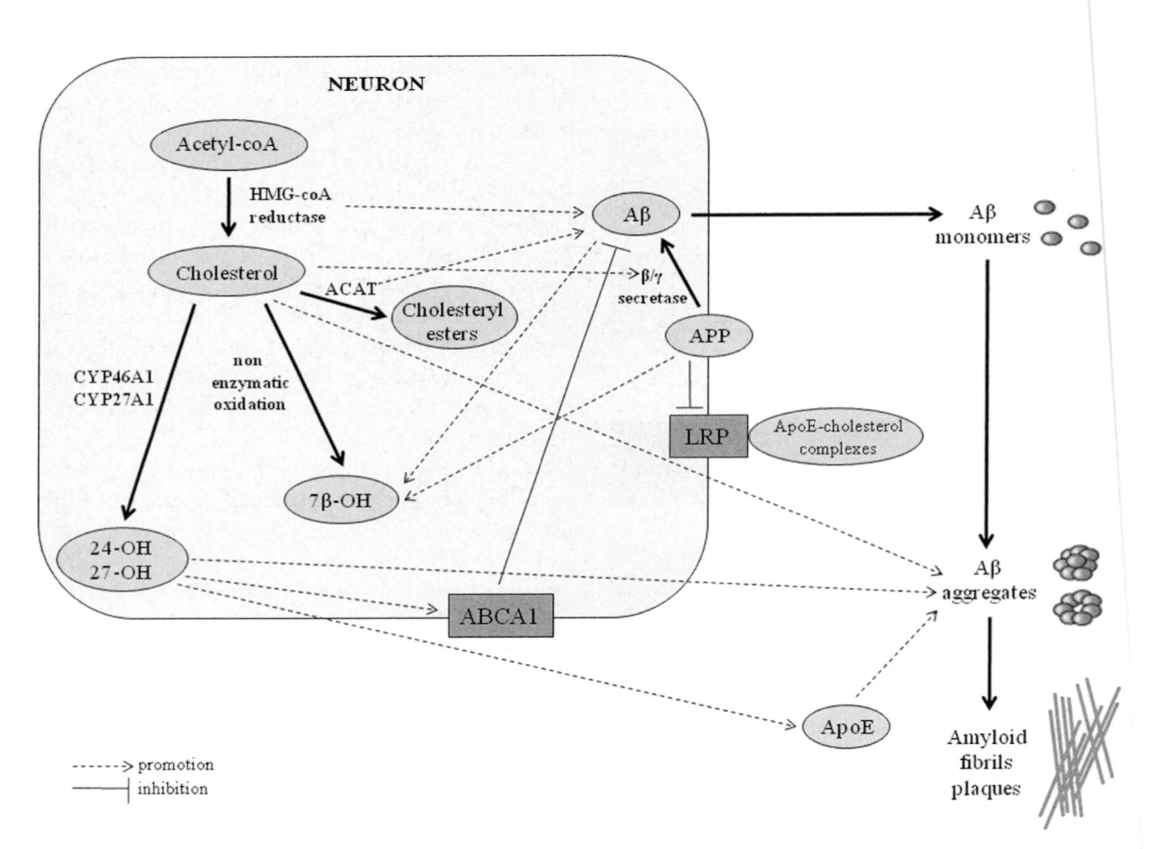

Figure 4. Interaction between lipid metabolism and amyloid-β (Aβ) processing. The major source of cerebral cholesterol is *de novo* synthesis from acetyl-coenzyme A mediated by 3-hydroxy-3-methyl-glutaryl-coenzyme A (HMG-CoA) reductase. Excess cholesterol is converted into cholesteryl esters by acyl-coenzyme A: cholesterol acyl-transferase (ACAT) and into the oxysterols 24- and 27-hydroxycholesterol (24-OH, 27-OH) by CYP47A1 and CYP27A1, respectively; the oxysterol 7β-hydroxycholesterol (7β-OH) may derive from cholesterol oxidation by Aβ and amyloid protein precursor (APP). Aβ production is increased by HMG-CoA reductase and ACAT activity and by cholesterol, which regulates β- and γ-secretase. LDL receptor–related protein (LRP) mediates the internalization of astrocyte-produced ApoE-cholesterol complexes into neurons. The intracellular domain of APP downregulates cellular cholesterol uptake by acting as a transcriptional suppressor of the LRP1 gene. 24-OH and 27-OH upregulate ATP-binding cassette transporter A1 (ABCA1), involved in cholesterol efflux, which modulates Aβ levels in neurons; these oxysterols increase ApoE levels and induce Aβ aggregation, which is also promoted by free ApoE and cholesterol.

In our recent study, we examined the ways in which the oxysterols 24-OH, 27-OH, and 7β-OH, specifically implicated in brain pathophysiology, may modulate and possibly amplify the expression of AD.[105] All three oxysterols strongly enhanced the binding and concentration of $A\beta_{42}$ on membranes of human differentiated neuronal cell lines (SK-N-BE and NT-2) by markedly upregulating expression and synthesis of CD36 and β1-integrin receptors, two components of the multireceptor complex CD36/β1-integrin/CD47, through which Aβ peptide binds to cell membrane.[88] An interesting finding of the same study is that only 24-OH significantly potentiates both the necrogenic and the apoptotic effects exerted by $A\beta_{42}$ peptide on these cells (Fig. 3A and B). These effects were inhibited when 24-OH-treated neuronal cells were incubated with anti-CD36 and anti-β1-integrin antibodies before $A\beta_{42}$ addition, since Aβ peptide binding to the cell surface was prevented. One significant reason for this selective behavior of 24-OH appears to be its marked prooxidant action on neuronal cells, by locally increasing ROS generation, an action not exerted by either 27-OH or 7β-OH (Fig. 3C). The 24-OH-dependent potentiation of Aβ neurotoxicity was completely inhibited by incubation of differentiated SK-N-BE or NT-2 cells with either the flavonol quercetin or the isoflavone genistein.[105]

However, despite the present knowledge demonstrates that cholesterol and some oxysterols play a key role in the AD pathogenesis interacting with Aβ (Fig. 4), elucidation of the precise mechanisms will clearly require further study.

Acknowledgments

The authors wish to thank the Italian Ministry of University, Prin 2008 and 2009, the Piedmontese Regional Government (Ricerca Sanitaria Finalizzata 2009), the CRT Foundation, Turin, and the University of Turin, Italy, for supporting this work.

Conflicts of interest

The authors declare no conflicts of interest.

References

1. Querfurth, H.W. & F.M. LaFerla. 2010. Alzheimer's disease. *N. Engl. J. Med.* **362:** 329–344.
2. Chopra, K., S. Misra & A. Kuhad. 2011. Neurobiological aspects of Alzheimer's disease. *Expert. Opin. Ther. Targets* **15:** 535–555.
3. Walsh, D.M. & D.J. Selkoe. 2007. A beta oligomers—a decade of discovery. *J. Neurochem.* **101:** 1172–1184.
4. Martins, I.J. *et al.* 2009. Cholesterol metabolism and transport in the pathogenesis of Alzheimer's disease. *J. Neurochem.* **111:** 1275–1308.
5. Carter, C.J. 2007. Convergence of genes implicated in Alzheimer's disease on the cerebral cholesterol shuttle: APP, cholesterol, lipoproteins, and atherosclerosis. *Neurochem. Int.* **50:** 12–38.
6. Wollmer, M.A. 2010. Cholesterol-related genes in Alzheimer's disease. *Biochim. Biophys. Acta* **1801:** 762–773.
7. Ghribi, O. 2008. Potential mechanisms linking cholesterol to Alzheimer's disease-like pathology in rabbit brain, hippocampal organotypic slices, and skeletal muscle. *J. Alzheimers Dis.* **15:** 673–684.
8. Di Paolo, G. & T.W. Kim. 2011. Linking lipids to Alzheimer's disease: cholesterol and beyond. *Nat. Rev. Neurosci.* **12:** 284–296.
9. Shobab, L.A., G.Y. Hsiung & H.H. Feldman. 2005. Cholesterol in Alzheimer's disease. *Lancet Neurol.* **4:** 841–852.
10. Panza, F. *et al.* 2006. Lipid metabolism in cognitive decline and dementia. *Brain Res. Rev.* **51:** 275–292.
11. Pappolla, M.A. *et al.* 2003. Mild hypercholesterolemia is an early risk factor for the development of Alzheimer amyloid pathology. *Neurology* **61:** 199–205.
12. Jick, H. *et al.* 2000. Statins and the risk of dementia. *Lancet* **356:** 1627–1631.
13. Wolozin, B. *et al.* 2000. Decreased prevalence of Alzheimer disease associated with 3-hydroxy-3-methyglutaryl coenzyme A reductase inhibitors. *Arch. Neurol.* **57:** 1439–1443.
14. Zamrini, E., G. McGwin & J.M. Roseman. 2004. Association between statin use and Alzheimer's disease. *Neuroepidemiology* **23:** 94–98.
15. Kandiah, N. & H.H. Feldman. 2009. Therapeutic potential of statins in Alzheimer's disease. *J. Neurol. Sci.* **283:** 230–234.
16. Fonseca, A.C. *et al.* 2010. Cholesterol and statins in Alzheimer's disease: current controversies. *Exp. Neurol.* **223:** 282–293.
17. Arvanitakis, Z. *et al.* 2008. Statins, incident Alzheimer disease, change in cognitive function, and neuropathology. *Neurology* **70:** 1795–1802.
18. Puglielli, L., R.E. Tanzi & D.M. Kovacs. 2003. Alzheimer's disease: the cholesterol connection. *Nat. Neurosci.* **6:** 345–351.
19. Evans, R.M. *et al.* 2004. Cholesterol and APOE genotype interact to influence Alzheimer disease progression. *Neurology* **62:** 1879–1881.
20. Bu, G. 2009. Apolipoprotein E and its receptors in Alzheimer's disease: pathways, pathogenesis and therapy. *Nat. Rev. Neurosci.* **10:** 333–344.
21. Kim, J., J.M. Basak & D.M. Holtzman. 2009. The role of apolipoprotein E in Alzheimer's disease. *Neuron* **63:** 287–303.
22. Marzolo, M.P. & G. Bu. 2009. Lipoprotein receptors and cholesterol in APP trafficking and proteolytic processing, implications for Alzheimer's disease. *Semin. Cell. Dev. Biol.* **20:** 191–200.
23. Jiang, Q. *et al.* 2008. ApoE promotes the proteolytic degradation of Abeta. *Neuron* **58:** 681–693.

24. Sagare, A. *et al.* 2007. Clearance of amyloid-beta by circulating lipoprotein receptors. *Nat. Med.* **13:** 1029–1031.
25. Cam, J.A. & G. Bu. 2006. Modulation of beta-amyloid precursor protein trafficking and processing by the low density lipoprotein receptor family. *Mol. Neurodegener.* **18:** 1–8.
26. Herz, J. & U. Beffert. 2000. Apolipoprotein E receptors: linking brain development and Alzheimer's disease. *Nat. Rev. Neurosci.* **1:** 51–58.
27. Liu, Q. *et al.* 2007. Amyloid precursor protein regulates brain apolipoprotein E and cholesterol metabolism through lipoprotein receptor LRP1. *Neuron* **56:** 66–78.
28. Jenner, A.M. *et al.* 2010. The effect of APOE genotype on brain levels of oxysterols in young and old human APOE epsilon2, epsilon3 and epsilon4 knock-in mice. *Neuroscience* **169:** 109–115.
29. Pfrieger, F.W. 2003. Cholesterol homeostasis and function in neurons of the central nervous system. *Cell. Mol. Life Sci.* **60:** 1158–1171.
30. Björkhem, I. & S. Meaney. 2004. Brain cholesterol: long secret life behind a barrier. *Arterioscler. Thromb. Vasc. Biol.* **24:** 806–815.
31. Pfrieger, F.W. 2003. Outsourcing in the brain: do neurons depend on cholesterol delivery by astrocytes? *Bioessays* **25:** 72–78.
32. Cordy, J.M., N.M. Hooper & A.J. Turner. 2006. The involvement of lipid rafts in Alzheimer's disease. *Mol. Membr. Biol.* **23:** 111–122.
33. Vetrivel, K.S. & G. Thinakaran. 2010. Membrane rafts in Alzheimer's disease beta-amyloid production. *Biochim. Biophys. Acta* **1801:** 860–867.
34. Beel, A.J. *et al.* 2010. Direct binding of cholesterol to the amyloid precursor protein: An important interaction in lipid-Alzheimer's disease relationships? *Biochim. Biophys. Acta* **1801**: 975–982.
35. Wahrle, S. *et al.* 2002. Cholesterol-dependent gamma-secretase activity in buoyant cholesterol-rich membrane microdomains. *Neurobiol. Dis.* **9:** 11–23.
36. Hur, J.Y. *et al.* 2008. Active gamma-secretase is localized to detergent-resistant membranes in human brain. *FEBS J.* **275:** 1174–1187.
37. Grimm, M.O. *et al.* 2008. Independent inhibition of Alzheimer disease beta- and gamma-secretase cleavage by lowered cholesterol levels. *J. Biol. Chem.* **283:** 11302–11311.
38. Xiong, H. *et al.* 2008. Cholesterol retention in Alzheimer's brain is responsible for high beta- and gamma-secretase activities and Abeta production. *Neurobiol. Dis.* **29:** 422–437.
39. Reid, P.C. *et al.* 2007. Alzheimer's disease: cholesterol, membrane rafts, isoprenoids and statins. *J. Cell. Mol. Med.* **11:** 383–392.
40. Harris, B., I. Pereira & E. Parkin. 2009. Targeting ADAM10 to lipid rafts in neuroblastoma SH-SY5Y cells impairs amyloidogenic processing of the amyloid precursor protein. *Brain Res.* **1296**: 203–215.
41. Yanagisawa, K. 2005. Cholesterol and amyloid beta fibrillogenesis. *Subcell. Biochem.* **38:** 179–202.
42. Kakio, A. *et al.* 2002. Interactions of amyloid beta-protein with various gangliosides in raft-like membranes: importance of GM1 ganglioside-bound form as an endogenous seed for Alzheimer amyloid. *Biochemistry* **41:** 7385–7390.
43. Harris, J.R. 2008. Cholesterol binding to amyloid-beta fibrils: a TEM study. *Micron* **9:** 1192–1196.
44. Kakio, A. *et al.* 2001. Cholesterol-dependent formation of GM1 ganglioside-bound amyloid beta-protein, an endogenous seed for Alzheimer amyloid. *J. Biol. Chem.* **276:** 24985–24990.
45. Burns, M.P. *et al.* 2006. Cholesterol distribution, not total levels, correlate with altered amyloid precursor protein processing in statin-treated mice. *Neuromolecular Med.* **8:** 319–328.
46. Puglielli, L. *et al.* 2001. Acyl-coenzyme A: cholesterol acyltransferase modulates the generation of the amyloid beta-peptide. *Nat. Cell Biol.* **3:** 905–912.
47. Bhattacharyya, R. & D.M. Kovacs. 2010. ACAT inhibition and amyloid beta reduction. *Biochim. Biophys. Acta* **1801:** 960–965.
48. Michikawa, M. *et al.* 2001. A novel action of alzheimer's amyloid beta-protein (Abeta): oligomeric Abeta promotes lipid release. *Neurosci.* **21:** 7226–7235.
49. Gong, J.S. *et al.* 2002. Amyloid beta-protein affects cholesterol metabolism in cultured neurons: implications for pivotal role of cholesterol in the amyloid cascade. *J. Neurosci. Res.* **70:** 438–446.
50. Fan, Q.W. *et al.* 2001. Cholesterol-dependent modulation of tau phosphorylation in cultured neurons. *J. Neurochem.* **76:** 391–400.
51. Koudinov, A.R. & N.V. Koudinova. 2005. Cholesterol homeostasis failure as a unifying cause of synaptic degeneration. *J. Neurol. Sci.* **229–230:** 233–240.
52. Mori, T. *et al.* 2001. Cholesterol accumulates in senile plaques of Alzheimer disease patients and in transgenic APP(SW) mice. *J. Neuropathol. Exp. Neurol.* **60:** 778–785.
53. Burns, M.P. *et al.* 2003. Co-localization of cholesterol, apolipoprotein E and fibrillar Abeta in amyloid plaques. *Brain Res. Mol. Brain Res.* **110:** 119–125.
54. Björkhem, I. *et al.* 2009. Oxysterols and neurodegenerative diseases. *Mol. Aspects Med.* **30:** 171–179.
55. Björkhem I. 2006. Crossing the barrier: oxysterols as cholesterol transporters and metabolic modulators in the brain. *J. Intern. Med.* **260:** 493–508.
56. Abildayeva, K. *et al.* 2006. 24(S)-hydroxycholesterol participates in a liver X receptor-controlled pathway in astrocytes that regulates apolipoprotein E-mediated cholesterol efflux. *J. Biol. Chem.* **281:** 12799–12808.
57. Meaney, S. *et al.* 2007. Novel route for elimination of brain oxysterols across the blood-brain barrier: conversion into 7alpha-hydroxy-3-oxo-4-cholestenoic acid. *J. Lipid Res.* **48:** 944–951.
58. Heverin, M. *et al.* 2005. Crossing the barrier: net flux of 27-hydroxycholesterol into the human brain. *J. Lipid Res.* **46:** 1047–1052.
59. Nelson, T.J. & D.L. Alkon. 2005. Oxidation of cholesterol by amyloid precursor protein and beta-amyloid peptide. *J. Biol. Chem.* **280:** 7377–7387.
60. Lütjohann, D. *et al.* 2000. Plasma 24S-hydroxycholesterol (cerebrosterol) is increased in Alzheimer and vascular demented patients. *J. Lipid Res.* **41:** 195–198.

61. Papassotiropoulos, A. *et al.* 2002. 24S-hydroxycholesterol in cerebrospinal fluid is elevated in early stages of dementia. *J. Psychiatr. Res.* **36:** 27–32.
62. Kölsch, H. *et al.* 2004. Altered levels of plasma 24S- and 27-hydroxycholesterol in demented patients. *Neurosci. Lett.* **368:** 303–308.
63. Bretillon, L. *et al.* 2000. Plasma levels of 24S-hydroxycholesterol in patients with neurological diseases. *Neurosci. Lett.* **293:** 87–90.
64. Bogdanovic, N. *et al.* 2001. On the turnover of brain cholesterol in patients with Alzheimer's disease. Abnormal induction of the cholesterol-catabolic enzyme CYP46 in glial cells. *Neurosci. Lett. Nov.* **314:** 45–48.
65. Brown, J. 3rd *et al.* 2004. Differential expression of cholesterol hydroxylases in Alzheimer's disease. *J. Biol. Chem.* **13:** 34674–34681.
66. Iuliano, L. *et al.* 2010. Vitamin E and enzymatic/oxidative stress-driven oxysterols in amnestic mild cognitive impairment subtypes and Alzheimer's disease. *J. Alzheimers Dis.* **21:** 1383–1392.
67. Heverin, M. *et al.* 2004. Changes in the levels of cerebral and extracerebral sterols in the brain of patients with Alzheimer's disease. *J. Lipid Res.* **45:** 186–193.
68. Shafaati, M. *et al.* 2011. Marked accumulation of 27-hydroxycholesterol in the brains of Alzheimer's patients with the Swedish APP 670/671 mutation. *J. Lipid Res.* **52:** 1004–1010.
69. Björkhem, I. *et al.* 2006. Oxysterols and Alzheimer's disease. *Acta Neurol. Scand. Suppl.* **185:** 43–49.
70. Leoni, V. *et al.* 2003. Side chain oxidized oxysterols in cerebrospinal fluid and the integrity of blood-brain and blood-cerebrospinal fluid barriers. *J. Lipid Res.* **44:** 793–799.
71. Yau, J.L. *et al.* 2003. Dehydroepiandrosterone 7-hydroxylase CYP7B: predominant expression in primate hippocampus and reduced expression in Alzheimer's disease. *Neuroscience* **121:** 307–314.
72. Leoni, V. *et al.* 2004. Diagnostic use of cerebral and extracerebral oxysterols. *Clin. Chem. Lab. Med.* **42:** 186–191.
73. Kim, W.S. *et al.* 2009. Impact of 27-hydroxycholesterol on amyloid-beta peptide production and ATP-binding cassette transporter expression in primary human neurons. *J. Alzheimers Dis.* **16:** 121–131.
74. Wang, L. *et al.* 2002. Liver X receptors in the central nervous system: from lipid homeostasis to neuronal degeneration. *Proc. Natl. Acad. Sci. USA* **15:** 13878–13883.
75. Riddell, D.R. *et al.* 2007. The LXR agonist TO901317 selectively lowers hippocampal Abeta42 and improves memory in the Tg2576 mouse model of Alzheimer's disease. *Mol. Cell. Neurosci.* **34:** 621–628.
76. Zelcer, N. *et al.* 2007. Attenuation of neuroinflammation and Alzheimer's disease pathology by liver x receptors. *Proc. Natl. Acad. Sci. USA* **19**: 10601–10606.
77. Cao, G. *et al.* 2007. Liver X receptor-mediated gene regulation and cholesterol homeostasis in brain: relevance to Alzheimer's disease therapeutics. *Curr. Alzheimer Res.* **4:** 179–184.
78. Famer, D. *et al.* 2007. Regulation of alpha- and beta-secretase activity by oxysterols: cerebrosterol stimulates processing of APP via the alpha-secretase pathway. *Biochem. Biophys. Res. Commun.* **359:** 46–50.
79. Prasanthi, J.R. *et al.* 2009. Differential effects of 24-hydroxycholesterol and 27-hydroxycholesterol on beta-amyloid precursor protein levels and processing in human neuroblastoma SH-SY5Y cells. *Mol. Neurodegener.* **4:** 1–8.
80. Marwarha, G. *et al.* 2010. Leptin reduces the accumulation of Abeta and phosphorylated tau induced by 27-hydroxycholesterol in rabbit organotypic slices. *J. Alzheimers Dis.* **19:** 1007–1019.
81. Sharma, S. *et al.* 2008. Hypercholesterolemia-induced Abeta accumulation in rabbit brain is associated with alteration in IGF-1 signaling. *Neurobiol. Dis.* **32:** 426–432.
82. Mateos, L. *et al.* 2009. Activity-regulated cytoskeleton-associated protein in rodent brain is down-regulated by high fat diet in vivo and by 27-hydroxycholesterol in vitro. *Brain Pathol.* **19:** 69–80.
83. Akiyama, H. *et al.* 2000. Inflammation and Alzheimer's disease. *Neurobiol. Aging* **21:** 383–421.
84. Guglielmotto, M. *et al.* 2010. Oxidative stress mediates the pathogenic effect of different Alzheimer's disease risk factors. *Front. Aging Neurosci.* **2:** 1–8.
85. Khandelwal, P.J., A.M. Herman & C.E. Moussa. 2011. Inflammation in the early stages of neurodegenerative pathology. *J. Neuroimmunol.* **15:** 1–11.
86. Sastre, M. *et al.* 2011. Inflammatory risk factors and pathologies associated with Alzheimer's disease. *Curr. Alzheimer Res.* **8:** 132–141.
87. Ferretti, M.T. & A.C. Cuello. 2011. Does a pro-inflammatory process precede Alzheimer's disease and mild cognitive impairment? *Curr. Alzheimer Res.* **8:** 164–174.
88. Verdier, Y., M. Zarándi & B. Penke. 2004. Amyloid beta-peptide interactions with neuronal and glial cell plasma membrane: binding sites and implications for Alzheimer's disease. *J. Pept. Sci.* **10:** 229–248.
89. Schwab, C. & P.L. McGeer. 2008. Inflammatory aspects of Alzheimer disease and other neurodegenerative disorders. *J. Alzheimers Dis.* **13:** 359–369.
90. Abbas, N. *et al.* 2002. Up-regulation of the inflammatory cytokines IFN-gamma and IL-12 and down-regulation of IL-4 in cerebral cortex regions of APP(SWE) transgenic mice. *J. Neuroimmunol.* **126:** 50–57.
91. Nunomura, A. *et al.* 2001. Oxidative damage is the earliest event in Alzheimer disease. *J. Neuropathol. Exp. Neurol.* **60:** 759–767.
92. Zhu, X. *et al.* 2007. Causes of oxidative stress in Alzheimer disease. *Cell. Mol. Life Sci.* **64:** 2202–2210.
93. Smith, M.A. *et al.* 2010. Increased iron and free radical generation in preclinical Alzheimer disease and mild cognitive impairment. *J. Alzheimers Dis.* **19:** 363–372.
94. Bonda, D.J. *et al.* 2010. Oxidative stress in Alzheimer disease: a possibility for prevention. *Neuropharmacology* **59:** 290–294.
95. Ding, Q., E. Dimayuga & J.N. Keller. 2007. Oxidative damage, protein synthesis, and protein degradation in Alzheimer's disease. *Curr. Alzheimer Res.* **4:** 73–79.
96. Praticò, D. *et al.* 2001. Increased lipid peroxidation precedes amyloid plaque formation in an animal model of Alzheimer amyloidosis. *J. Neurosci.* **21:** 4183–4187.

97. Naylor, R., A.F. Hill & K.J. Barnham. 2008. Neurotoxicity in Alzheimer's disease: is covalently crosslinked A beta responsible? *Eur. Biophys. J.* **37:** 265–268.
98. Arca, M. *et al.* 2007. Increased plasma levels of oxysterols, in vivo markers of oxidative stress, in patients with familial combined hyperlipidemia: reduction during atorvastatin and fenofibrate therapy. *Free Radic. Biol. Med.* **42:** 698–705.
99. Melov, S. *et al.* 2007. Mitochondrial oxidative stress causes hyperphosphorylation of tau. *PLoS One* **2:** e536.
100. Cassano, T. *et al.* 2011. Glutamatergic alterations and mitochondrial impairment in a murine model of Alzheimer disease. *Neurobiol. Aging* **27.** doi: 10.1016!j.neurobiolaging.2011.09.021.
101. Usui, K. *et al.* 2009. Site-specific modification of Alzheimer's peptides by cholesterol oxidation products enhances aggregation energetics and neurotoxicity. *Proc. Natl. Acad. Sci. USA* **106:** 18563–18568.
102. Kolsch, H. *et al.* 2000. 7alpha-Hydroperoxycholesterol causes CNS neuronal cell death. *Neurochem. Int.* **36:** 507–512.
103. Ong, W.Y. *et al.* 2010. Changes in brain cholesterol metabolome after excitotoxicity. *Mol. Neurobiol.* **41:** 299–313.
104. Ferrera, P. *et al.* 2008. Cholesterol potentiates beta-amyloid-induced toxicity in human neuroblastoma cells: involvement of oxidative stress. *Neurochem. Res.* **33:** 1509–1517.
105. Gamba, P. *et al.* 2011. Interaction between 24-hydroxycholesterol, oxidative stress, and amyloid-β in amplifying neuronal damage in Alzheimer's disease: three partners in crime. *Aging Cell* **10:** 403–417.

Ann. N.Y. Acad. Sci. ISSN 0077-8923

ANNALS OF THE NEW YORK ACADEMY OF SCIENCES
Issue: *Environmental Stressors in Biology and Medicine*

Rottlerin and curcumin: a comparative analysis

Emanuela Maioli,[1] Claudia Torricelli,[1] and Giuseppe Valacchi[2,3]

[1]Department of Physiology, University of Siena, Siena, Italy. [2]Department of Biology and Evolution, University of Ferrara, Ferrara, Italy. [3]Department of Food and Nutrition, Kyung Hee University, Seoul, South Korea

Address for correspondence: Emanuela Maioli, Department of Physiology, Via Aldo Moro, 7-53100 Siena, Italy. emanuela.maioli@unisi.it

Rottlerin and curcumin are natural plant polyphenols with a long tradition in folk medicine. Over the past two decades, curcumin has been extensively investigated, while rottlerin has received much less attention, in part, as a consequence of its reputation as a selective PKCδ inhibitor. A comparative analysis of genomic, proteomic, and cell signaling studies revealed that rottlerin and curcumin share a number of targets and have overlapping effects on many biological processes. Both molecules, indeed, modulate the activity and/or expression of several enzymes (PKCδ, heme oxygenase, DNA methyltransferase, cyclooxygenase, lipoxygenase) and transcription factors (NF-κB, STAT), and prevent aggregation of different amyloid precursors (α-synuclein, amyloid Aβ, prion proteins, lysozyme), thereby exhibiting convergent antioxidant, anti-inflammatory, and antiamyloid actions. Like curcumin, rottlerin could be a promising candidate in the fight against a variety of human diseases.

Keywords: Rottlerin; curcumin; antiamyloid; antioxidant; anti-inflammatory

Introduction

For centuries, many plant compounds have had an outstanding role in medicine and are still used either directly or after chemical modification. Phytochemicals, on the basis of their structure can be grouped into different classes, the most common of which are the polyphenols, characterized by multiple hydroxylated aromatic rings. Polyphenols are very abundant in nature and extremely diverse. Given their great heterogeneity, it is impossible to define a structure–activity relationship, and a more appropriate approach is to analyze their biological effects and the associated signal-transduction pathways.

One of the most intriguing and most extensively investigated polyphenols is curcumin (Fig. 1, upper panel), which has been the subject of hundreds of published papers over the past two decades. Curcumin has a wide range of pharmacological activities, which include antitumor, antioxidant, antiamyloid, and anti-inflammatory properties.[1]

Rottlerin (Fig. 1, lower panel) has received much less attention than curcumin. Rottlerin came to attention in 1994 when it was identified as a selective PKCδ inhibitor in a publication by Gschwendt *et al.*[2] As a consequence of this designation, this molecule has long been neglected, despite its potential, though sporadic, off-site targeting activity.[3,4] Only recently has investigation of the properties of this molecule shifted beyond PKCδ inhibition; several PKCδ-independent rottlerin effects have been described so far.[5–7]

The limited availability of rottlerin has also likely hampered interest in this molecule and, in the past, discouraged research of its medicinal potential. Rottlerin, unlike other polyphenols, is not present in edible vegetables and in common beverages; instead, it is primarily present in the gland hair covering the fruit of *Mallotus philippinensis* (Euphorbiaceae), an evergreen rain forest tree that is inedible and only used by indigenous populations of Southeast Asian tropical regions. Rottlerin is used as a dye for coloring textiles and as an old folk remedy against tapeworm (when taken orally) and scabies and ringworm (when administered topically).

Similarly, curcumin, a polyphenol derived from the spice turmeric, has a long tradition as a drug against a variety of diseases, including fever,

doi: 10.1111/j.1749-6632.2012.06514.x

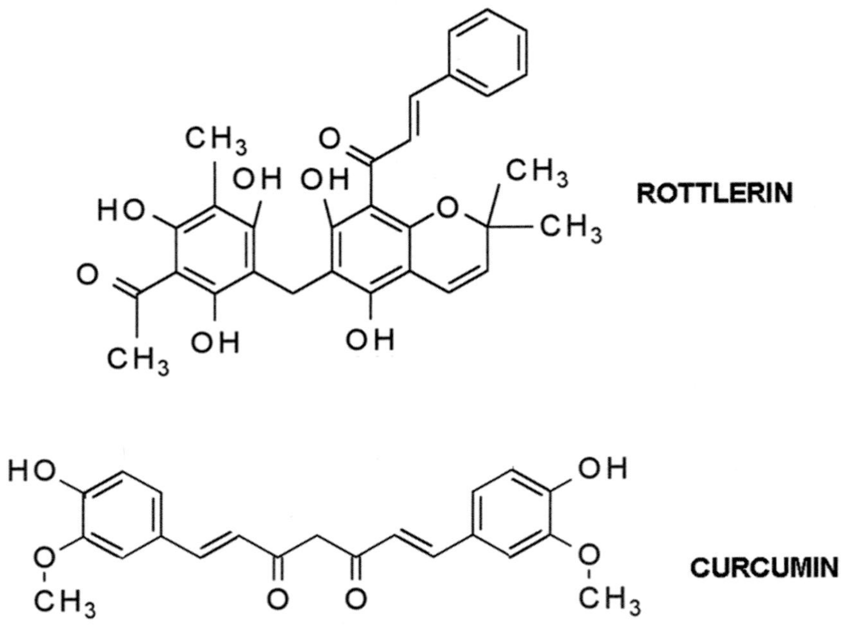

ROTTLERIN

CURCUMIN

Figure 1. The chemical structures of rottlerin and curcumin.

bronchitis, parasitic worms, and bladder and kidney inflammations. However, unlike rottlerin, a number of rigorous scientific studies of curcumin, mainly conducted by the Aggarwal group, have been available since 1994,[8] and even earlier by other groups.[9]

A comparative analysis of data derived from genomic, proteomic, and cell signaling studies revealed that these two molecules share a number of targets and have convergent effects on many biological processes. In this short review, we will present some of the most documented and/or tempting findings concerning the pharmacological activities that curcumin and rottlerin have in common, with the aim of launching the latter as a promising candidate in the fight against a variety of human diseases.

PKCδ inhibition

As mentioned earlier, rottlerin was first used as a tool to investigate PKCδ signaling, and much of what is known about the role of PKCδ in biological processes has been derived from such studies.

Although it is known that rottlerin does not exhibit inhibitory effect on PKCδ activity *in vitro*,[10] and its use as a blocker of PKCδ activity has been discouraged,[11] several studies support the concept that rottlerin does inhibit PKCδ translocation and activity in cell cultures, and there can be more than one explanation for this apparent controversy.[12] Rottlerin is a mitochondrial uncoupler that, by lowering ATP levels, can prevent PKCδ tyrosine phosphorylation and activation.[4] In addition, rottlerin can cause PKCδ cleavage via caspase-3 activation, thereby preventing PKCδ membrane translocation and signaling.[12] These indirect inhibitory effects, as well as the eventual lack of inhibition, are likely cell specific.

Curiously, though investigation of the roles of curcumin and rottlerin in PKCδ targeting began roughly at the same time, its evolution over the years has been profoundly different. Indeed, the concept of PKC as a curcumin target dates back to 1993.[8,13,14] In early reports, curcumin was described as an inhibitor of PKCs, PKCδ included. Recent studies reported that curcumin decreases PKCδ protein levels in the mouse neuroblastoma cell line, neuro 2A, by forming tight complexes with the enzyme and generating tyrosine phosphorylated PKCδ fragments.[15] The same decrease in PKCδ protein levels has been also observed in human hepatocellular carcinoma Hep 3B cells, but has been ascribed to PKCδ gene downregulation.[16]

The physical interaction of curcumin (and derivatives) with the diacylglycerol (DAG)- and phorbol ester–sensitive C1B subdomain of novel PKCs, has been very recently investigated by Majhi *et al.*

using fluorescence spectroscopy. They found that by forming hydrogen bonds with the activator binding domain of the enzyme, curcumin and its analogues could change PKC(δ) conformation and influence activation and membrane translocation properties.[17] A similar study also indicated that both carbonyl and hydroxyl groups of curcumin are important for PKCs binding.[18] It is not clear from these last studies if curcumin acts as an activator or inhibitor of PKC, and further work is needed to definitively establish the impact of curcumin on enzyme activity.

Antioxidant properties

The antioxidant properties of crude extracts of the *Mallotus philippinensis* fruit have been known since 2007,[19] but the evidence that pure rottlerin acts as a radical scavenger was presented by Longpre *et al.* in 2008.[20] In this study, empirically established concentrations of rottlerin and *N,N*-diphenyl-*N*-picrylhydrazyl (DPPH) revealed that the ability of rottlerin to scavenge the stable free radical DPPH was comparable to curcumin.

More recently, the antioxidant properties of rottlerin have been confirmed by our group.[21] The reaction between rottlerin and DPPH was measured by UV and electron paramagnetic resonance spectroscopy, and we determined that two molecules of DPPH are reduced by one molecule of rottlerin, indicating that rottlerin can donate two hydrogens from the five CH2-bond phenolic hydroxyl groups. Subsequently, the antioxidant properties of rottlerin were also confirmed in cultured cells (MCF-7). According to Longpre's study, showing that rottlerin prevented deoxycholate-induced reactive oxygen species (ROS) generation in human colon epithelial cells, it was demonstrated that rottlerin neutralized hydrogen peroxide added to the MCF-7 cells and prevented free radical generation.

That curcumin has antioxidant activity is widely known. The chemical structure of curcumin is peculiar; it has two o-methoxyl phenolic OH groups attached to the α,β-unsaturated β-diketone (heptadiene-dione) moiety (Fig. 1, upper panel). The free radical scavenging activity can arise either from the phenolic OH group or from the methylene CH_2 group of the β-diketone moiety.

Although there has been some debate on the two different sites free radical attack, based on the most current experimental and theoretical results, it can be concluded that the phenolic OH plays a major role in the scavenging ability of curcumin.[22] Also the impact of the curcumin scavenger activity *in vivo* (cultured cells) has been extensively demonstrated.[23] However, it should be noted that curcumin also shows pro-oxidant effects, depending on the dose, the chemical context, and the assay used. Specifically, curcumin can generate ROS in the presence of transition metals, such as Cu^2, thereby causing copper-dependent DNA damage.[24] This curcumin pro-oxidant action has been primarily ascribed to the β-diketone moiety.[25]

Moreover, both rottlerin and curcumin are able to modulate heme oxygenase (HO)-1, a cytoprotective enzyme that plays critical roles in resistance to oxidative stress. A recent report by Park *et al.*[26] demonstrated that rottlerin induces HO-1 expression at both the protein and messenger level in HT29 cells, and this effect was accompanied by the production of ROS, activation of p38 and ERK, and nuclear translocation of nuclear erythroid 2 p45–related factor 2 (Nrf2), a major redox-sensitive transcription factor that regulates expression of several antioxidant defense genes. To our knowledge, this is the only report describing an increase in ROS levels after rottlerin treatment and conflicts with the findings reported earlier.[19–21] Further work is needed to verify the anti- and/or pro-oxidant action of Rottlerin *in vivo* (cultured cells).

Suppression of PKCδ expression by siRNA or overexpression of WT-PKCδ did not abrogate the rottlerin-mediated induction of HO-1, suggesting that rottlerin works through PKCδ-independent pathways in HT29 cells.[26] In addition, Park *et al.*[26] reported that rottlerin increased HO-1 protein in lung cancer, renal carcinoma, colon cancer, and hepatocellular carcinoma cells, suggesting that HO-1 upregulation, regardless of the mechanism, is a common response to rottlerin in a variety of tumor cells. Nevertheless, some of the findings obtained in this interesting study are in apparent contrast with other reports, where rottlerin, through PKCδ blockage, acts as an inhibitor of HO-1 expression induced by cigarette smoke or acrolein.[27,28] In another report, rottlerin, again used as a PKCδ inhibitor, was able to suppress HO-1 induction by the phytoestrogen puerarin.[29]

It should be noted, however, that the work mentioned earlier[27–29] was performed on noncancer cells (human bronchial epithelial cells, mouse brain endothelial cells, and mouse mesangial cells, respectively), and the authors did not compare the combined treatment with the effect of rottlerin alone.

In addition, the HO-1 gene is under the control of several transcription factors, which can be differently engaged by the different stimuli.[30] Therefore, the discrepancy may be due to both cellular specificity and signaling pathways, in addition to the nature of the stimulus.

In contrast, the literature on curcumin is much more coherent. In fact, all published papers describe curcumin as an inductor of HO-1 in cancer as well as in noncancer cells. An early work reported that curcumin, at low concentrations, markedly upregulated HO-1 levels in endothelial cells, whereas at higher concentrations, a gradual decline in HO-1 activity was observed.[31] HO-1 upregulation occurred only after prolonged treatment (18 h) and no effect was observed after a 1.5 h exposure. Although NF-κB and AP-1 are known regulators of HO-1 gene expression, the authors excluded their involvement because curcumin is a potent inhibitor of both transcription factors. The authors also excluded that the curcumin induction of HO-1 was due to pro-oxidant action, because the addition of the antioxidant and thiol donor *N*-acetylcysteine did not affect the increase in HO-1 activity by the drug.

Conversely, in human hepatoma cells, HO-1 induction by curcumin was ascribed to ROS generation, and pretreatment with *N*-acetylcysteine prevented HO-1 induction by curcumin.[32] In these cancer cells, HO-1 was induced via activation of Nrf2 and p38 and phosphatase inhibition. A similar mechanism has been described in vascular smooth muscle cells, where induction of HO-1 by curcumin occurred through translocation of Nrf2 into the nucleus and activation of antioxidant response element.[33] No study, however, has examined curcumin in association with other treatments (such as cigarette smoke, acrolein, etc.), likely because curcumin is now widely accepted as an inductor of HO-1. There is also work in which curcumin and rottlerin were used together in human monocytes.[34] In this study, rottlerin prevented curcumin-induced HO-1 mRNA expression by inhibiting PKCδ, but the effect of rottlerin alone on HO-1 expression was not investigated.

In summary, while it is clear that rottlerin and curcumin, directly or indirectly, positively or negatively, affect HO-1 expression, a comparative mechanistic analysis between the drugs is problematic because of the marked differences among research protocols and cell models.

Epigenetic effects

Epigenetic effects refer to reversible alterations that modify gene expression, without changing the DNA sequence. DNA methylation is perhaps the most studied epigenetic modification. This process occurs through addition of a methyl group at the 5 position of the cytosine residues. In humans, DNA methylation is catalyzed by enzymes called DNA methyltransferases (DNMT) in the presence of *S*-adenosyl-methionine, as a methyl donor.[35]

Methylation typically occurs in the cytosine–phosphate–guanine (CpG) dinucleotides of DNA. CpGs are often grouped in clusters, the CpG islands, which are found overlapping with or near gene promoters. In normal tissues, most CpG islands are unmethylated. During normal physiological processes and in many diseases, such as cancer, CpG islands can however be methylated, and this chemical modification interferes with transcription factor binding and generally leads to gene silencing.

The effect of curcumin on the promoter methylation of different genes has been evaluated by independent research groups. In mouse prostate cancer cells, curcumin is able to restore the epigenetically silenced (hypermethylated) Nrf2 gene, a master regulator of major antioxidative stress enzymes such as the previously mentioned HO-1[33] and NAD(P)H dehydrogenase quinone 1 (NQO-1).[36]

Similarly, in human prostate cancer cells, curcumin causes hypomethylation of the Neurog1 gene *(NERUOG1)* promoter, thereby restoring the expression of this gene, which is considered a cancer-related CpG-methylation epigenome marker.[37] A hypomethylating effect of curcumin has also been described in cervical cancer cell lines by Jha *et al.*,[38] who found that treatment with curcumin led to hypomethlylation and activation of the tumor suppressor retinoic acid receptor β2 (*RARβ2*) gene. However, in lung cancer cells and tissues, other groups have found that demethoxycurcumin and bisdemethoxycurcumin, but not curcumin, hypomethylated the tumor suppressor gene Wnt inhibitory factor-1 (WIF-1).[39]

In addition, there are reports demonstrating a minimal effect of curcumin on DNMTs mRNA and protein levels. The authors concluded that the hypomethylation observed was not caused by DNMT downregulation but rather by direct inhibition of DNMT enzymatic activity by curcumin.[36,37]

In a recent study in a leukemia cell line, Liu *et al.*[40] showed that curcumin induced global DNA hypomethylation. Molecular docking studies on the interaction of curcumin and the DNMT1 homology model indicated that curcumin exerts an inhibitory effect by covalently blocking the catalytic thiolate of C1226 of DNMT1. Shortly thereafter, however, an independent group concluded that curcumin had little or no pharmacologically relevant activity as a DNMT inhibitor, using a virtual screening of a large database of natural products with a validated homology model of the catalytic domain of DNMT1.[41]

The question of whether curcumin inhibits the expression or the activity of DNMTs is still under debate. It is important to note that reconciling the conflicting findings derived from different models and experimental and theoretical studies is not easy. Nevertheless, in this case, the majority of studies demonstrate that curcumin is a potent hypomethylating agent, either by modulating the activity or the expression of DNMT enzymes.

To our knowledge, there are only two papers reporting that rottlerin decreased DNMT mRNA and protein expression. In the first study, it was demonstrated that rottlerin-induced promoter hypomethylation, downregulated DNMT (both DNMT1 and DNMT3A), and consequently upregulated the cell cycle inhibitors p16INK4A and p21WAF1 in human colon cancer cells.[42] In another study, rottlerin (as well as the 5-aza-2′-deoxycytidine DNMT inhibitor, ERK-MAPK inhibitors, and MEK1/2 siRNA transfection) decreased DNMT1 expression (not DNMT3A or DNMT3B) and increased p16INK4A and p21WAF1 protein levels in the same colon cancer cells.[43]

In the studies mentioned earlier, the epigenetic effects of rottlerin were ascribed to selective PKCδ inhibition and blockage of the signaling pathway to mitogen-activated protein kinase (MAPK), thus making a comparative analysis with curcumin particularly problematic.

Despite the limited experimental evidence and mechanistic information regarding the epigenetic effects of rottlerin, it is tempting to speculate that both rottlerin and curcumin have the ability to reverse CpG island hypermethylation of various methylation-silenced genes (mainly tumor suppressor genes) and to reactivate expression (Fig. 2). Consistently, it has been demonstrated that both rottlerin[12] and curcumin[1] possess anticancer properties.

Although further investigation is needed to definitively establish the impact of curcumin,

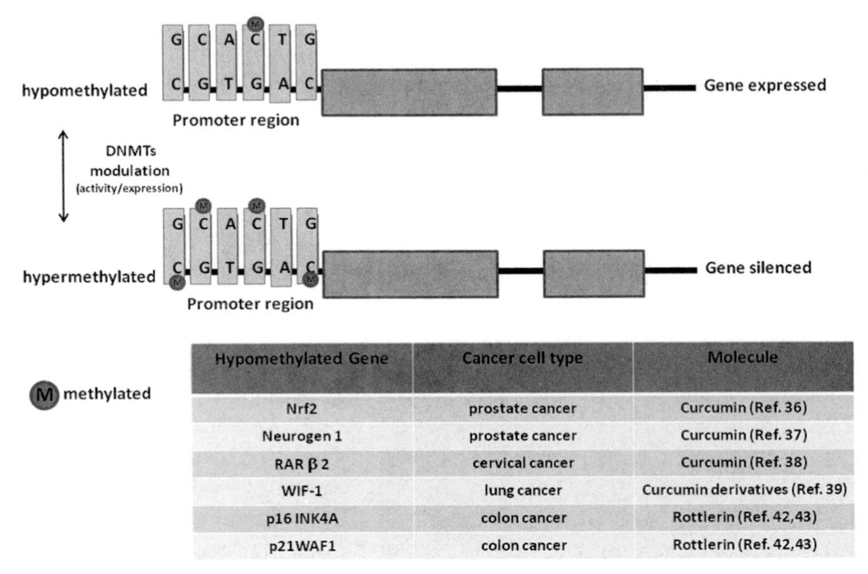

Hypomethylated Gene	Cancer cell type	Molecule
Nrf2	prostate cancer	Curcumin (Ref. 36)
Neurogen 1	prostate cancer	Curcumin (Ref. 37)
RAR β 2	cervical cancer	Curcumin (Ref. 38)
WIF-1	lung cancer	Curcumin derivatives (Ref. 39)
p16 INK4A	colon cancer	Rottlerin (Ref. 42,43)
p21WAF1	colon cancer	Rottlerin (Ref. 42,43)

Figure 2. The epigenetic effects of rottlerin and curcumin through modulation of DNMTs activity. See the section "Epigenetic effects" for more details.

and especially of rottlerin, in epigenetic gene modulation, the knowledge accumulated so far encourages further study of these natural compounds in relation to cancer and other diseases.

Antiamyloid properties

The history of rottlerin and amyloidosis began with the observation that many chemical protein kinases inhibitors, which are believed to compete with the ATP binding site of the kinases, actually inhibit a number of proteins that do not bind to ATP.

In an early study by McGovern *et al.*,[44] a panel of widely used compounds at micromolar concentrations, including rottlerin and quercetin, was found to inhibit three diverse nonkinase enzymes (β-lactamase, chymotrypsin, and malate dehydrogenase) with similar potency. Specifically, rottlerin inhibited β-lactamase with an IC50 of 1.2 μM.

Dynamic light scattering experiments showed that all of these inhibitors formed promiscuous aggregates in aqueous solutions, suggesting that the aggregates rather than the monomeric form may physically sequester enzymes, thereby nonspecifically inhibiting them.

Moreover, Coan *et al.* showed that the aggregate–protein interaction (e.g., β-lactamase-rottlerin complex) resulted in partial protein denaturation and increased susceptibility to proteolytic digestion, thus offering an additional explanation for enzyme inhibition.[45]

Protein denaturation could have several other biological implications. For example, it suggests a possible molecular mechanism to explain the induction of endoplasmic reticulum (ER) stress observed in cells treated with rottlerin.[46] In fact, the accumulation of unfolded or misfolded proteins activates a set of transmembrane ER stress sensors and triggers the so-called unfolded protein response, a process that can lead to the activation of the apoptotic pathway.[47]

Of note, the same phenomenon (ER stress and apoptosis) has also been observed in curcumin-treated cells.[48] Because of their hydrophobic nature, rottlerin, and curcumin, both freely pass cellular membranes and could likely accumulate inside cytoplasm and organelles.[49]

Furthermore, the propensity of many small hydrophobic polyphenols to form promiscuous aggregates in aqueous solutions may also have an effect on the absorption of these drugs from the intestinal fluids.[50] It is well known that a major drawback of curcumin is its poor oral bioavailability, although mainly ascribed to water insolubility and low stability in the gastrointestinal tract, and may additionally result from colloidal aggregation.

In reference to amyloid inhibition, the link between promiscuous aggregates and antifibrillogenic activity was provided by Feng *et al.*,[51] who demonstrated that several known chemical aggregators inhibit fibrillation of amyloid precursors in biochemical assays. Conversely, some known amyloid inhibitors form colloidal aggregates, suggesting that protein sequestration into colloidal particles can serve as an inhibitory mechanism for both enzyme activity and amyloid fibrillation.

As described earlier, rottlerin has both aggregation-based inhibitory activity toward enzymes and antiamyloid properties. Indeed, rottlerin has been shown to inhibit amyloid formation of the yeast and mouse prion proteins Sup35 and recMoPrP.[51] Further, Sarkar *et al.* reported that rottlerin prevented protein aggregation into highly ordered fibrils of two model proteins belonging to different classes, that is, lysozyme and cytochrome c. Molecular docking results indicate that rottlerin binds in the vicinity of amyloidogenic regions of both proteins, thereby blocking their aggregation.[52]

The same group recently performed a detailed study on the antiamyloidogenic properties of rottlerin. By monitoring thioflavin T and employing anisotropy measurements, the authors showed that in rottlerin, the disaggregation effect is faster (five min) compared to other known compounds, such as curcumin (one to two days).[53]

Despite the sporadic and limited number of studies, the literature suggests that rottlerin can function as a general antiamyloid agent. In fact, regardless of the sequence of amyloid-forming proteins, the process of aggregation is similar in all cases, and the resulting materials exhibit common structural motifs and tinctorial characteristics.

Although curcumin has never been investigated as an aggregation-based inhibitor, a wealth of studies *in vitro* and *in vivo* substantiate the claim that curcumin has antiamyloid properties. The curcumin structure resembles that of the colloidal inhibitors, an emblematic example is that of Congo red dye, one of the first molecules demonstrated to exhibit colloidal inhibition.[54] In fact, curcumin and Congo red share a similar chemical scaffold: two substituted aromatic groups separated by a rigid, planar backbone.[55]

Moreover, curcumin and curcuminoids are known to directly bind (and inhibit) various intracellular and extracellular proteins and enzymes, with high affinity *in vitro*, through a combination of hydrophobic interactions and hydrogen bonds.[56] This notion and the fact that circumin is poorly absorbed in the intestine suggest that it could belong to the family of compounds that have a great propensity to self-aggregate in aqueous media. This behavior, if verified, could also explain, in part, the pleiotropic activities of curcumin (and rottlerin).

As an antiamyloid, curcumin has been shown to exert antifibrillogenic activity and disaggregating properties against the fibrillation/aggregation of various amyloid proteins, such as α-synuclein present in Parkinson's disease[57] and the β-amyloid peptide present in Alzeimer's disease.[58–61] Curcumin inhibits amyloid Aβ1–42 oligomer formation and cell toxicity at micromolar concentrations *in vitro* and binds to plaques reducing amyloid levels *in vivo*.[62]

Recent studies, employing surface plasmon resonance experiments, revealed that curcumin-decorated nanosized liposomes showed enhanced binding affinity for Aβ1–42 fibrils, with very low dissociation rate constants that are likely due to multivalent interactions that make the binding between curcumin and Aβ1–42 fibrils nearly irreversible.[63]

The antiamyloid property of curcumin is not limited to Aβ1–42 and α-synuclein but also extends to islet amyloid polypeptides,[64] prion proteins present in Creutzfeldt-Jakob's disease,[65] and lysozymes.[66] Like rottlerin,[53] curcumin inhibits lysozyme fibrillation and also induces disaggregation of existing lysozyme fibrils. Moreover, preincubated curcumin improved inhibitory activity against lysozyme fibrillation compared with fresh curcumin. Notably, this behavior is considered a hallmark of promiscuous inhibitors.[44] Interestingly, the observed enhanced inhibitory potency was associated with formation of dimeric curcumin species (possible aggregates) during pre-incubation under certain conditions.

In summary, these results suggest that rottlerin, like many other small hydrophobic molecules, may serve as a prototype for the development of novel therapeutics against different human amyloidoses (Fig. 3).

Regulation of the inflammatory response

Some phloroglucinol compounds, isolated from the pericarps of Mallotus Japonicus, were reported to modulate the activity of macrophages and monocytes, exhibiting a series of anti-inflammatory actions, such as inhibition of NO

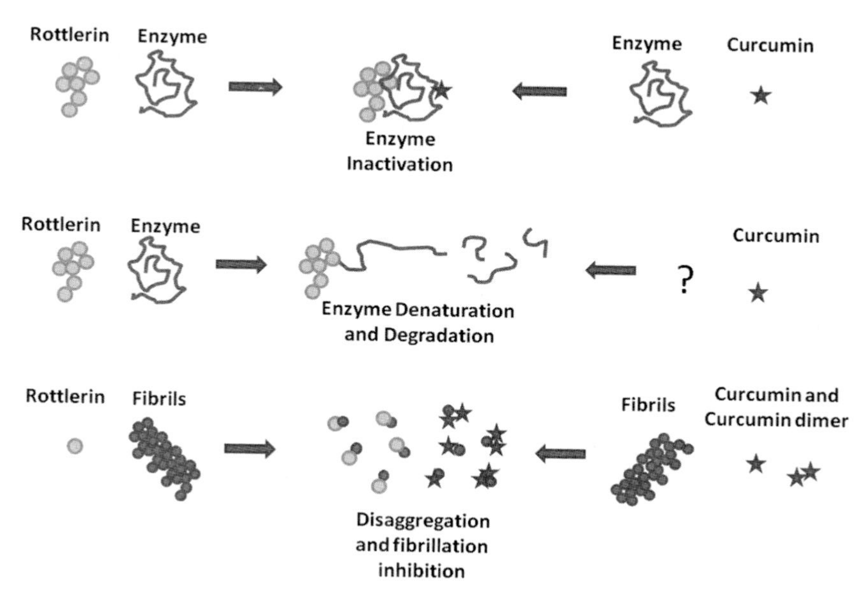

Figure 3. Rottlerin and curcumin inhibitory activity toward enzymes and amyloid precursors. See the section "Antiamyloid properties" for more details.

release,[67] suppression of PGE2 production, and downregulation of proinflammatory cytokines.[68] In particular, two rottlerin-like compounds, isomallotochromanol and isomallotochromene, exhibited the most potent effect. Interestingly, these phloroglucinol derivatives have been shown to inhibit NF-κB activation and transcriptional activity, thereby causing downregulation of NF-κB target gene products such as TNF-α and IL-6. Accordingly, the phloroglucinol rottlerin, purified from *Mallotus philippinensis*, has been reported to possess anti-inflammatory effects, for instance through inhibition of arachidonic acid pathways, which are discussed later.

Cyclooxygenase (COX) is the rate-limiting enzyme for prostaglandin synthesis, catalyzing the conversion of arachidonic acid to prostaglandins and related eicosanoids. Two important forms of COX have been identified: the constitutively expressed COX-1 and the inducible COX-2 isoform, which is expressed only under inflammatory stimulation. The COX-2 promoters include CRE, AP-1, NFIL-6, and NF-κB *cis*-acting elements.

In a study of endothelial cells, rottlerin decreased acrolein-induced COX-2 at both the protein and messenger levels,[69] and a similar inhibitory effect has been observed in astrocytes treated with the inflammatory mediator Bradykinin.[70]

Rottlerin also inhibited manganese-induced COX-2 expression in A549 human lung epithelial cells,[71] and COX-2 expression and PGE2 biosynthesis induced by the 5-lipoxygenase inhibitor Zileuton in cardiac myogenic H9c2 cells.[72] In all of these studies, rottlerin was used as a PKCδ inhibitor and the observed effects were ascribed to PKCδ blockage.

Moreover, in this same PKCδ-dependent way, rottlerin inhibited thrombin-induced COX-2 expression in vascular smooth muscle cells.[73] Curcumin was also used and found to prevent thrombin-induced COX-2 in smooth muscle cells, but through inhibition of AP-1 and NF-κB.

Although, as stated here and elsewhere, whether rottlerin inhibiting PKCδ cannot be ruled out, it is, however, worth mentioning that rottlerin, like curcumin, has been demonstrated to prevent NF-κB nuclear migration and transcriptional activity in several cell types.[21,74–76] Thus it could reasonably downregulate the NF-κB target gene COX-2 through a PKCδ-independent pathway.

To further complicate the picture, the bulk of the literature indicates that rottlerin has anti-inflammatory properties via PKCδ inhibition. In MDA-MB-231 breast cancer cells, rottlerin enhanced IL-1β–induced COX-2 expression at both the protein and mRNA level, and sustained activation of p38 MAPK in a PKCδ-independent manner.[77] Although we are not able to explain these opposite findings, which are derived from different protocols and cellular models, we noted that MDA-MB-231 cells express the PKCδ isoform at very low levels with respect to other cell types, and this could explain the discrepancy.[78]

Rottlerin has also been reported to modulate 15-lipoxygenase (15-LO) expression. Lipoxygenases, like COX, are inflammatory enzymes implicated in the arachidonic acid metabolism that catalyzes the insertion of oxygen into various positions in arachidonic acid, resulting in the production of leukotrienes. In two studies where rottlerin was used to explore the involvement of PKCδ in the serine phosphorylation of signal transducer and activator of transcription (STAT) proteins in IL-13–stimulated primary human monocytes, the authors reported that pretreatment with rottlerin inhibited IL-13–induced 15-LO mRNA and protein expression.[79,80] This effect was ascribed to PKCδ inhibition, because PKCδ silencing evoked the same response. Mechanistically, rottlerin inhibited PKC δ-mediated phosphorylation of STAT3 in Ser 727, which thereby inhibited the IL-13–induced STAT3 binding to DNA.

The anti-inflammatory potential of curcumin was first described by Aggarwal and coworkers, who demonstrated that curcumin prevents the degradation of the inhibitory protein IκBα and the subsequent nuclear translocation of the p65 subunit.[81] Since then, a number of laboratories have demonstrated the inhibitory effects of curcumin on both COX-2 and LOX, and various mechanisms of action have been described in addition to NF-κB inhibition (for reviews, see Refs. 82 and 83).

It is worth stressing, however, that curcumin, while decreasing COX and LOX expression, not only acts at the transcriptional level, likely via multiple mechanisms (NF-κB, STAT, Ap-1), but also complexes with and inhibits enzyme activity. COX and LOX indeed belong to the list of proteins directly targeted by curcumin.[56] It may be important to identify

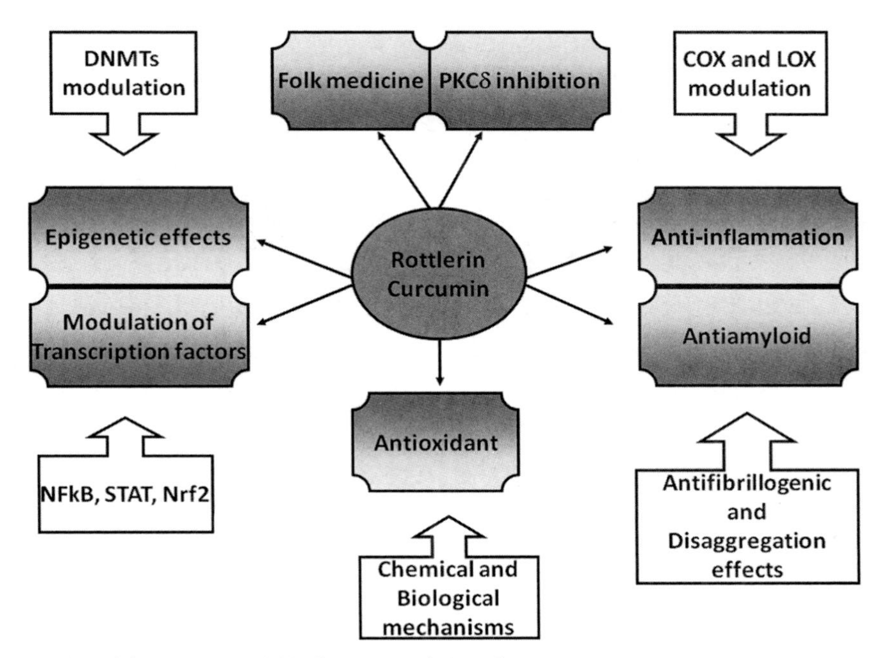

Figure 4. An overview of the common activities between rottlerin and curcumin.

whether rottlerin, as a promiscuous inhibitor, is able to directly inhibit COX and LOX.

We speculate the existence of common signaling molecules and pathways that rottlerin and curcumin share in executing their anti-inflammatory effects. For instance, independent studies have found that curcumin decreased phorbol ester-induced COX-2 protein expression and PGE2 production.[84–86] Because phorbol esters are known to activate classical and novel (PKCδ) PKCs, it is possible that curcumin, like rottlerin, downregulated COX-2 via PKC(δ) inhibition.[87]

In addition, Kim *et al.*[88] demonstrated that curcumin suppressed the ganglioside-, lipopolysaccharide-, or interferon γ–stimulated induction of COX-2 through inhibition of Janus kinase (JAK)–STAT signaling in the rat brain. Interestingly, the curcumin mechanism of action resembles that of rottlerin in the suppression of IL-13–induced 15-LOX via inhibition of STAT3 phosphorylation in human monocytes.[79,80]

In summary, a plausible conclusion is that both rottlerin and curcumin possess anti-inflammatory properties by inhibiting arachidonic acid metabolism through some common pathways, such as interference with PKCδ signaling, inhibition of NF-κB and STAT transcriptional activity, and/or direct binding to COX and LOX.

Concluding remarks

This comparative analysis suggests that rottlerin, although much less studied, possesses some pharmacological properties that overlap with those of the more extensively investigated curcumin. Both of these low-molecular weight, natural polyphenols have a wide range of medicinal potential, including antitumor, antioxidant, antiamyloid, and anti-inflammatory activities (Fig. 4). Other common effects, not discussed in this paper, include mitochondrial uncoupling action,[4,89] antibacterial activity,[90,91] and inhibitory effects on the mTOR pathway.[92,93]

A remarkable difference between curcumin and rottlerin can be recognized in their relative bioavailability. In fact, animal studies have shown that curcumin is rapidly metabolized, conjugated in the liver, and excreted in the feces, and therefore possesses limited bioavailability and low systemic distribution. In contrast, in an animal model of Parkinson's (mice), rottlerin, administered both intraperitoneally (3–7 mg/Kg) and orally (20 mg/Kg), exhibited protective effects and was not toxic. Most importantly, HPLC measurements revealed that rottlerin reached the target tissues (brain) in an intact and active form (1125 pg/mg brain tissue).[94]

To conclude, we believe that the results presented in this short review, of the pleiotropic activities of rottlerin and curcumin, are strong motivation for future studies to assess the real potential of these promising molecules in the treatment of several human diseases.

Conflicts of interest

The authors declare no conflicts of interest.

References

1. Maheshwari, R.K. *et al.* 2006. Multiple biological activities of curcumin: a short review. *J. Life Sci.* **78:** 2081–2087.
2. Gschwendt, M. *et al.* 1994. Rottlerin, a novel protein kinase inhibitor. *Biochem. Biophys. Res. Commun.* **199:** 63–98.
3. Bain, J. *et al.* 2007. The selectivity of protein kinase inhibitors: a further update. *Biochem. J.* **408:** 297–315.
4. Soltoff, S.P. 2001. Rottlerin is a mitochondrial uncoupler that decreases cellular ATP levels and indirectly blocks protein kinase Cdelta tyrosine phosphorylation. *J. Biol. Chem.* **276:** 37986–37992.
5. Tillman, D.M. *et al.* 2003. Rottlerin sensitizes colon carcinoma cells to tumor necrosis factor-related apoptosis-inducing ligand-induced apoptosis via uncoupling of the mitochondria independent of protein kinase C. *Cancer Res.* **63:** 5118–5125.
6. Xu, S.Z. 2007. Rottlerin induces calcium influx and protein degradation in cultured lenses independent of effects on protein kinase C delta. *Basic Clin. Pharmacol. Toxicol.* **101:** 459–464.
7. Tapia, J.A., R.T. Jensen & L.J. García-Marín. 2006. Rottlerin inhibits stimulated enzymatic secretion and several intracellular signaling transduction pathways in pancreatic acinar cells by a non-PKC-delta-dependent mechanism. Biochim. *Biophys. Acta* **1763:** 25–38.
8. Reddy, S. & B.B. Aggarwal. 1994. Curcumin is a non-competitive and selective inhibitor of phosphorylase kinase. *FEBS Lett.* **341:** 19–22.
9. Kuttan, R. *et al.* 1985. Potential anticancer activity of turmeric (Curcuma longa). *Cancer Lett.* **29:** 197–202.
10. Davies, S.P. *et al.* 2000. Specificity and mechanism of action of some commonly used protein kinase inhibitors. *Biochem. J.* **351:** 95–105.
11. Soltoff, S.P. 2007. Rottlerin: an inappropriate and ineffective inhibitor of PKCdelta. *Trends Pharmacol. Sci.* **28:** 453–458.
12. Maioli, E., C. Torricelli & G. Valacchi 2011. Rottlerin and cancer: novel evidences and mechanisms. *Scientific World Journal.* doi:10.1100/2012/350826
13. Liu, J.Y., S.J. Lin & J.K. Lin. 1993. Inhibitory effects of curcumin on protein kinase C activity induced by 12-O-tetradecanoyl-phorbol-13-acetate in NIH 3T3 cells. *Carcinogenesis* **14:** 857–861.
14. Hasmeda, M. & G.M. Polya. 1996. Inhibition of cyclic AMP-dependent protein kinase by curcumin. *Phytochemistry* **42:** 599–605.
15. Conboy, L. *et al.* 2009. Curcumin-induced degradation of PKC delta is associated with enhanced dentate NCAM PSA expression and spatial learning in adult and aged Wistar rats. *Biochem. Pharmacol.* **77:** 1254–1265.
16. Kao, H.H. *et al.* 2011. Kinase gene expression and subcellular protein expression pattern of protein kinase C isoforms in curcumin-treated human hepatocellular carcinoma Hep 3B cells. *Plant Foods Hum. Nutr.* **66:** 136–142.
17. Majhi, A. *et al.* 2010. Binding of curcumin and its long chain derivatives to the activator binding domain of novel protein kinase C. *Bioorg. Med. Chem.* **18:** 1591–1598.
18. Das, J. *et al.* 2011. Binding of isoxazole and pyrazole derivatives of curcumin with the activator binding domain of novel protein kinase C. *Bioorg. Med. Chem.* **19:** 6196–6202.
19. Arfan, M. *et al.* 2007. Antioxidant activity of extracts of Mallotus philippinensis fruit and bark. *J. Food Lipids* **14:** 280–297.
20. Longpre, J.M. & G. Loo. 2008. Protection of human colon epithelial cells against deoxycholate by rottlerin. *Apoptosis* **13:** 1162–1171.
21. Maioli, E. *et al.* 2009. Rottlerin inhibits ROS formation and prevents NFkappaB activation in MCF-7 and HT-29 cells. *J. Biomed. Biotechnol.* [Epub ahead of print.]
22. Priyadarsini, K.I. *et al.* 2003. Role of phenolic O-H and methylene hydrogen on the free radical reactions and antioxidant activity of curcumin. *Free Radic. Biol. Med.* **35:** 475–484.
23. Feng, J.Y. & Z.Q. Liu. 2009. Phenolic and enolic hydroxyl groups in curcumin: which plays the major role in scavenging radicals? *J. Agric. Food Chem.* **57:** 11041–11046.
24. Ahsan, H. *et al.* 1999. Pro-oxidant, anti-oxidant and cleavage activities on DNA of curcumin and its derivatives demethoxycurcumin and bisdemethoxycurcumin. *Chem. Biol. Interact.* **121:** 161–175.
25. Yoshino, M. *et al.* 2004. Prooxidant activity of curcumin: copper-dependent formation of 8-hydroxy-2'-deoxyguanosine in DNA and induction of apoptotic cell death. *Toxicol. In vitro* **18:** 783–789.
26. Park, E.J. *et al.* 2010. Rottlerin induces heme oxygenase-1 (HO-1) up-regulation through reactive oxygen species (ROS) dependent and PKC d-independent pathway in human colon cancer HT29 cells. *Biochimie* **92:** 110–115.
27. Zhang, H. & H.J. Forman. 2008. Acrolein induces heme oxygenase-1 through PKC-delta and PI3K in human bronchial epithelial cells. *Am. J. Respir. Cell. Mol. Biol.* **38:** 483–490.
28. Shih, R.H. *et al.* 2011. Cigarette smoke extract up-regulates heme oxygenase-1 via PKC/NADPH oxidase/ROS/PDGFR/PI3K/Akt pathway in mouse brain endothelial cells. *J Neuroinflammation* **8:** 104.
29. Kim, K.M. *et al.* 2010. Puerarin suppresses AGEs-induced inflammation in mouse mesangial cells: a possible pathway through the induction of heme oxygenase-1 expression. *Toxicol. Appl. Pharmacol.* **244:** 106–113.
30. Sikorski, E.M. *et al.* 2004. The story so far: Molecular regulation of the heme oxygenase-1 gene in renal injury. *Am. J. Physiol. Renal Physiol.* **286:** 425–441.

31. Motterlini, R. *et al.* 2000. Curcumin, an antioxidant and anti-inflammatory agent, induces heme oxygenase-1 and protects endothelial cells against oxidative stress. *Free Radic. Biol. Med.* **28:** 1303–1312.
32. McNally, S.J. *et al.* 2007. Curcumin induces heme oxygenase 1 through generation of reactive oxygen species. *Int. J. Mol. Med.* **19:** 165–172.
33. Pae, H.O. *et al.* 2007. Roles of heme oxygenase-1 in curcumin-induced growth inhibition in rat smooth muscle cells. *Exp. Mol. Med.* **39:** 267–277.
34. Rushworth, S.A. *et al.* 2006. Role of protein kinase C delta in curcumin-induced antioxidant response element-mediated gene expression in human monocytes. *Biochem. Biophys. Res. Commun.* **341:** 1007–1016.
35. Herman, J.G. & S.B. Baylin. 2003. Gene silencing in cancer in association with promoter hypermethylation. *N. Engl. J. Med.* **349:** 2042–2054.
36. Khor, T.O. *et al.* 2011. Pharmacodynamics of curcumin as DNA hypomethylation agent in restoring the expression of Nrf2 via promoter CpGs demethylation. *Biochem. Pharmacol.* **82:** 1073–1078.
37. Shu, L. *et al.* 2011. Epigenetic CpG demethylation of the promoter and reactivation of the expression of neurog1 by curcumin in prostate LNCaP cells. *AAPS J.* **13:** 606–614. doi:10.1208/s12248-011-9300-y.
38. Jha, A.K. *et al.* 2010. Reversal of hypermethylation and reactivation of the RARβ2 gene by natural compounds in cervical cancer cell lines. *Folia Biol. (Praha).* **56:** 195–200.
39. Liu, Y.L. *et al.* 2011. Hypomethylation effects of curcumin, demethoxycurcumin and bisdemethoxycurcumin on WIF-1 promoter in non-small cell lung cancer cell lines. *Mol. Med. Report.* **4:** 675–679.
40. Liu, Z. *et al.* 2009. Curcumin is a potent DNA hypomethylation agent. *Bioorg. Med. Chem. Lett.* **19:** 706–709.
41. Medina-Franco, J.L. *et al.* 2011. Natural products as DNA methyltransferase inhibitors: a computer-aided discovery approach. *Mol. Divers.* **15:** 293–304.
42. Chen, Z.F. *et al.* 2006. The effect of PKC-delta inhibitor Rottlerin on human colon cancer cell line SW1116 and its mechanism. *Zhonghua Zhong Liu Za Zhi.* **28**: 564–567.
43. Lu, R. *et al.* 2007. Inhibition of the extracellular signal-regulated kinase/mitogen-activated protein kinase pathway decreases DNA methylation in colon cancer cells. *J. Biol. Chem.* **282:** 12249–12259.
44. McGovern, S.L. & B.K. Shoichet. 2003. Kinase inhibitors: not just for Kinases anymore. *J. Med. Chem.* **46:** 1478–1483.
45. Coan, K.E. *et al.* 2009. Promiscuous aggregate-based inhibitors promote enzyme unfolding. *J. Med. Chem.* **52:** 2067–2075.
46. Lim, J.H. *et al.* 2008. Rottlerin induces pro-apoptotic endoplasmic reticulum stress through the protein kinase C-delta-independent pathway in human colon cancer cells. *Apoptosis* **13:** 1378–1385.
47. Lai, E., T. Teodoro & A. Volchuk. 2007. Endoplasmic reticulum stress: signaling the unfolded protein response. *Physiology (Bethesda)* **22:** 193–201.
48. Wu, S.H. *et al.* 2010. Curcumin induces apoptosis in human non-small cell lung cancer NCI-H460 cells through ER stress and caspase cascade- and mitochondria-dependent pathways. *Anticancer Res.* **30:** 2125–2133.
49. Wahlang, B., Y.B. Pawar & A.K. Bansal. 2011. Identification of permeability-related hurdles in oral delivery of curcumin using the Caco-2 cell model. *Eur. J. Pharm. Biopharm.* **77:** 275–282.
50. Doak, A.K. *et al.* Colloid formation by drugs in simulated intestinal fluid. *J. Med. Chem.* **53:** 4259–4265.
51. Feng, B.Y. *et al.* 2008. Small-molecule aggregates inhibit amyloid polymerization. *Nat. Chem. Biol.* **4:** 197–199.
52. Sarkar, N., M. Kumar & V.K. Dubey. 2011. Exploring possibility of promiscuity of amyloid inhibitor: studies on effect of selected compounds on folding and amyloid formation of proteins. *Process Biochem.* **46:** 1179–1185.
53. Sarkar, N., M. Kumar & V.K. Dubey. 2011. Rottlerin dissolves pre-formed protein amyloid: a study on hen egg white lysozyme. *Biochim. Biophys. Acta.* **1810:** 809–814.
54. McGovern, S.L. *et al.* 2002. A common mechanism underlying promiscuous inhibitors from virtual and high-throughput screening. *J. Med. Chem.* **45:** 1712–1722.
55. Reinke, A.A. & J.E. Gestwicki. 2007. Structure–activity relationships of amyloid Beta-aggregation inhibitors based on curcumin: influence of linker length and flexibility. *Chem. Biol. Drug Des.* **70:** 206–215.
56. Subash, C. *et al.* 2011. Multitargeting by curcumin as revealed by molecular interaction studies. *Nat. Prod. Rep.* **28:** 1937–1955. doi:10.1039/c1np00051a.
57. Pandey, N. *et al.* 2008. Curcumin inhibits aggregation of alpha-synuclein. *Acta Neuropathol.* **115:** 479–489.
58. Ono, K. *et al.* 2004. Curcumin has potent anti-amyloidogenic effects for Alzheimer's beta-amyloid fibrils *in vitro*. *J. Neurosci. Res.* **75:** 742–750.
59. Kim, H. *et al.* 2005. Effects of naturally occurring compounds on fibril formation and oxidative stress of beta-amyloid. *J. Agric. Food Chem.* **53:** 8537–8541.
60. Kim, D.S., S.Y. Park & J.K. Kim. 2001. Curcuminoids from curcuma longa L. (zingiberaceae) that protect PC12 rat pheochromocytoma and normal human umbilical vein endothelial cells from abeta(1–42) insult. *Neurosci. Lett.* **303:** 57–61.
61. Re, F. *et al.* 2010. Beta amyloid aggregation Inhibitors: small molecules as candidate drugs for therapy of Alzheimer's disease. *Curr. Med. Chem.* **17:** 2990–3006.
62. Yang, F. *et al.* 2005. Curcumin inhibits formation of amyloid beta oligomers and fibrils, binds plaques, and reduces amyloid *in vivo*. *J. Biol. Chem.* **280:** 5892–5901.
63. Mourtas, S. *et al.* 2011. Curcumin-decorated nanoliposomes with very high affinity for amyloid-β1–42 peptide. *Biomaterials.* **32:** 1635–1645.
64. Daval, M. *et al.* 2010. The effect of curcumin on human islet amyloid polypeptide misfolding and toxicity. *Amyloid* **17:** 118–128.
65. Hafner-Bratkovic, I. *et al.* 2008. Curcumin binds to the alpha-helical intermediate and to the amyloid form of prion protein—a new mechanism for the inhibition of PrP(Sc) accumulation. *J. Neurochem.* **104:** 1553–1564.
66. Steven, S.S. *et al.* 2009. Effect of curcumin on the amyloid fibrillogenesis of hen egg-white lysozyme. *Biophys. Chem.* **144:** 78–87.

67. Ishii, R. *et al.* 2002. Inhibitory effects of phloroglucinol derivatives from Mallotus japonicus on nitric oxide production by a murine macrophage-like cell line, RAW 264.7, activated by lipopolysaccharide and interferon-gamma. *Biochim. Biophys. Acta.* **15:** 74–82.
68. Ishii, R. *et al.* 2002. Prostaglandin E(2) production and induction of prostaglandin endoperoxide synthase-2 is inhibited in a murine macrophage-like cell line, RAW 264.7, by Mallotus japonicus phloroglucinol derivatives. *Biochim. Biophys. Acta.* **1571:** 115–123.
69. Park, Y.S. *et al.* 2007. Acrolein induces Cyclooxygenase-2 and prostaglandin production in human umbilical vein endothelial cells: roles of p38 MAP Kinase. *Arterioscler. Thromb. Vasc. Biol.* **27:** 1319–1325.
70. Hsieh, H.L. *et al.* 2007. BK-induced COX-2 expression via PKC-δ-dependent activation of p42/p44 MAPK and NF-κB in astrocytes. *Cell. Signal.* **19:** 330–340.
71. Jang, B.C. 2009. Induction of COX-2 in human airway cells by manganese: role of PI3K/PKB, p38 MAPK, PKCs, Src, and glutathione depletion. *Toxicol. In vitro* **23:** 120–126.
72. Kwak, H.J. *et al.* 2010. The cardioprotective effects of zileuton, a 5-lipoxygenase inhibitor, are mediated by COX-2 via activation of PKCδ. *Cell. Signal.* **22:** 80–87.
73. Hsieh, H.L. *et al.* 2008. PKC-δ/c-Src-mediated EGF receptor transactivation regulates thrombin-induced COX-2 expression and PGE2 production in rat vascular smooth muscle cells. *Biochim. Biophys. Acta* **1783:** 1563–1575.
74. Torricelli, C. *et al.* 2008. Rottlerin inhibits the nuclear factor kappaB/cyclin-D1 cascade in MCF-7 breast cancer cells. *Life Sci.* **82:** 638–643.
75. Valacchi, G. *et al.* 2009. Rottlerin: a multifaced regulator of keratinocyte cell cycle. *Exp. Dermatol.* **18:** 516–521.
76. Valacchi, G. *et al.* 2011. Rottlerin exhibits antiangiogenic effects *in vitro. Chem. Biol. Drug Des.* **77:** 460–470.
77. Park, E.J. & T.K. Kwon. 2011. Rottlerin enhances IL-1b-induced COX-2 expression through sustained p38 MAPK activation in MDA-MB-231 human breast cancer cells. *Exp. Mol. Med.* **4:** 669–675.
78. Shanmugam, M. *et al.* 2001. A role for protein kinase C delta in the differential sensitivity of MCF-7 and MDA-MB 231 human breast cancer cells to phorbol ester-induced growth arrest and p21(WAFI/CIP1) induction. *Cancer Lett.* **172:** 43–53.
79. Xu, B. *et al.* 2004. Role of Protein Kinase C isoforms in the regulation of Interleukin-13-induced 15-Lipoxygenase gene expression in human monocytes. *J. Biol. Chem.* **279:** 15954–15960.
80. Bhattacharjee, A. *et al.* 2006. Monocyte 15-Lipoxygenase expression is regulated by a novel cytosolic signaling complex with Protein Kinase C d and tyrosine-phosphorylated Stat3. *J. Immunol.* **177:** 3771–3781.
81. Singh, S. & B.B. Aggarwal. 1995. Activation of transcription factor NF-kappa B is suppressed by curcumin (diferuloylmethane). *J. Biol. Chem.* **270:** 24995–25000.
82. Rao, C.V. 2007. Regulation of COX and LOX by curcumin. *Adv. Exp. Med. Biol.* **595:** 213–226.
83. Menon, V.P. & A.R. Sudheer. 2007. Antioxidant and anti-inflammatory properties of curcumin. *Adv. Exp. Med. Biol.* **595:** 105–125.
84. Zhang, F. *et al.* 1999. Inhibition of cyclo-oxygenase 2 expression in colon cells by the chemopreventive agent curcumin involves inhibition of NF-kappaB activation via the NIK/IKK signaling complex. *Carcinogenesis* **20:** 445–451.
85. Zhang, F. *et al.* 1999. Curcumin inhibits cyclooxygenase-2 transcription in bile acid- and phorbol ester-treated human gastrointestinal epithelial cells. *Carcinogenesis.* **20:** 445–451.
86. Ireson, C. *et al.* 2001. Characterization of metabolites of the chemopreventive agent curcumin in human and rat hepatocytes and in the rat *in vivo*, and evaluation of their ability to inhibit phorbol ester-induced prostaglandin E2 production. *Cancer Res.* **61:** 1058–1064.
87. Lin, J.K. 2004. Suppression of protein kinase C and nuclear oncogene expression as possible action mechanisms of cancer chemoprevention by Curcumin. *Arch. Pharm. Res.* **27:** 683–692.
88. Kim H.Y. *et al.* 2003. Curcumin suppresses Janus kinase-STAT inflammatory signaling through activation of Src homology 2 domain-containing tyrosine phosphatase 2 in brain microglia. *J. Immunol.* **171:** 6072–6079.
89. Lim, H.W., H.Y. Lim & K.P. Wong. 2009. Uncoupling of oxidative phosphorylation by curcumin: implication of its cellular mechanism of action. *Biochem. Biophys. Res. Commun.* **389:** 187–192.
90. Zaidi, S.F.H. *et al.* 2009. Potent bactericidal constituents from Mallotus philippinensis against clarithromycin and metronidazole resistant strains of Japanese and Pakistani Helicobacter pylori. *Biol. Pharm. Bull.* **32:** 631–636.
91. De, R. *et al.* 2009. Antimicrobial activity of curcumin against Indian Helicobacter pylori and also during mice infection. *Antimicrob. Agents Chemother.* **53:** 1592–1597.
92. Balgi, A.D. *et al.* 2009. Screen for chemical modulators of autophagy reveals novel therapeutic inhibitors of mTORC1 signaling. *PLoS One* **4:** e7124.
93. Yu, S. *et al.* 2008. Curcumin inhibits Akt/mammalian target of rapamycin signaling through protein phosphatase dependent mechanism. *Mol. Cancer Ther.* **7:** 2609–2620.
94. Zhang, D. *et al.* 2007. Neuroprotective effect of protein kinase Cdelta inhibitor rottlerin in cell culture and animal models of Parkinson's disease. *J. Pharmacol. Exp. Ther.* **322:** 913–922.

Ann. N.Y. Acad. Sci. ISSN 0077-8923

ANNALS OF THE NEW YORK ACADEMY OF SCIENCES
Issue: *Environmental Stressors in Biology and Medicine*

Plant polyphenols and human skin: friends or foes

Liudmila Korkina, Chiara De Luca, and Saveria Pastore
Laboratory of Tissue Engineering and Skin Pathophysiology, Dermatology Institute (IDI IRCCS), Rome, Italy

Address for correspondence: Liudmila Korkina, Laboratory of Tissue Engineering and Skin Pathophysiology, Dermatology Institute (IDI IRCCS), Via Monti di Creta 104, Rome 00167, Italy. l.korkina@idi.it

In response to abiotic and biotic stressors, numerous polyphenols (PPs) are synthesized from phenylalanine by higher plants, amid many other plants that are poisonous for insects, birds, animals, and humans. PPs are also widely recognized by botanical dermatology as major plant constituents inducing allergic reactions, contact dermatitis, phytodermatoses, and photophytodermatoses. Notwithstanding these clinical observations, thousands of cosmetic/dermatological preparations based on PP-containing plant extracts or pure PPs emerge yearly with the claims of photoprotection, chemoprevention of skin tumors, anti-aging, wound healing, etc. However, because of their peculiar physical, chemical, and biological properties, PPs could be a double-edged sword for human skin, exerting both protective and damaging actions. Here, we distinguish direct and indirect anti- and pro-oxidant properties of PPs, their interactions with major xenobiotic metabolic systems and sensors/receptors of environmental hazards, anti- and proinflammatory potential, and photoprotection versus photosensitization.

Keywords: human skin; inflammation; plant polyphenols; UV irradiation; xenobiotic metabolism

Introduction

There is a steadily growing interest in the use of natural products to prevent and combat environment-associated skin pathologies, such as contact dermatitis, skin allergies, tumors, bacterial, and viral infections, as well as premature skin aging. In this sense, plant-derived polyphenols (PPs) are the most discussed substances as potential anti-inflammatory,[1–4] cancer preventive,[5,6] photoprotective,[7,8] and antibacterial remedies.[9] PPs represent active components of plant extracts, which have been widely used in Western and traditional medicine to treat several chronic skin diseases, such as psoriasis[10,11] and vitiligo.[12,13] They are also known to accelerate skin wound healing[14,15] and to possess anti-inflammatory effects when applied topically.[2,9] They are believed to protect human skin against deleterious effects of solar irradiation, ozone, and other airborne toxins.[16–18] On the other hand, there have been enduring clinical observations that plants can cause a variety of skin pathologies, such as contact, allergic, and phytophotodermatitis,[19] and can induce skin pigmentation defects due to increased or suppressed melanogenesis.[19] Apart from unintentional or occupational contact with skin-damaging plants, there are numerous cosmetic preparations, skin care products, and food supplements containing plant extracts or dried plant materials that could lead to undesirable skin defects in sensitive individuals.[20,21] The clinical symptoms from contact, such as skin irritation, inflammation, ulceration, and enhanced/suppressed pigmentation, are mainly attributed to PPs and their metabolites.[22,23] Mammals have evolved a great number of detoxifying metabolic pathways in the skin, gastrointestinal tract, and other epithelial linings at risk of contact with plant components to quickly eliminate these potentially toxic natural substances.[22–24]

Here, we discuss the molecular mechanisms of plant PPs/their metabolites interaction with human skin components leading to opposite, healing, or harmful (esthetic or anti-esthetic) outcomes: anti- and pro-oxidant properties, activation versus inhibition of the skin metabolic system, anti- and proinflammatory effects, and photoprotective and phototoxic action.

doi: 10.1111/j.1749-6632.2012.06510.x

PPs as chemical and biological anti-/pro-oxidants

Polyphenols are widely distributed in nature and in the plant kingdom in particular. PPs are synthesized by higher plants in response to environmental hazards that induce enhanced free-radical production in various plant parts.[25] Thus, plant exposure to UV light, gamma irradiation, ozone, low temperatures, organic toxins, and heavy metals results in the induction of phenylalanine ammonia lyase (PAL, EC 4-3.1-5), which catalytically deaminates phenylalanine to cinnamic acid.[26] This reaction is the key step in polyphenol biosynthesis. The parent molecules, (cinnamic acid and its close metabolites) are further catalytically transformed to a large variety of PPs:[22,26] glycosylated phenylpropanoids, flavonoids and isoflavonoids, stilbenoids, coumarines, curcuminoids, and phenolic polymers, such as tannins (synonimous to proanthocyanidins), suberin, lignins, and lignans. The superfamily of PPs (more than 4,000 substances have been described so far[27]) is composed of structurally and functionally diverse and, nevertheless, biologically similar active substances. Their similarity is determined by the unique parent molecule and by the presence of phenoxyl groups, which largely determine their peculiar chemical and photochemical reactivity.

The great majority of publications consider direct (chemical) free-radical scavenging/antioxidant effects of PPs as a molecular basis for their beneficial biological effects.[22,25,28,29] Notwithstanding structural diversity, PPs usually possess low-redox potentials ($E^{\circ} = 0.25$–0.75V) and easily donate one electron to compounds with higher redox potentials, such as free radicals that result in free-radical scavenging:

$$\mathrm{Ph-OH + R^{*} \rightarrow Ph-O^{*} + RH}, \qquad (1)$$

where Ph-OH, Ph-O* and R* are the PP, the phenoxyl radical, and a free radical, respectively.[30–32] The phenoxyl radical may react with a second free radical:

$$\mathrm{Ph-O^{*} + R^{*} \rightarrow RH + Ph-quinone}, \qquad (2)$$

where Ph-quinone is a stable quinone.

The high reactivity of PPs and Ph-O* toward R* initiating free radical–driven chain reactions underlies the chain-breaking mechanism of antioxidant activity of PPs. Moreover, effective superoxide scavenging by PPs results in the inhibiton of peroxynitrite formation, which suppresses nitrosative stress:[33]

$$\mathrm{^{*}O_2 - + {}^{*}NO + H+ \rightarrow HOONO}. \qquad (3)$$

One of the possible precursors of PPs transformation in the skin is through interaction with nitric oxide (NO)–derived redox-active substances.[34] As a result of this interaction, PPs are converted into toxic nitrated metabolites.

Alternatively, in the presence of molecular oxygen as it occurs on the skin surface, plant PPs with low-redox potentials, such as epigallocatechin gallate (EGCG), quercetin, myricetin, and many others, may be easily oxidized through a free radical–driven chain autoxidation process converting the PP to a quinone structure during the propagation stage:

$$\mathrm{RH_3^{*} + O_2 \rightarrow RH_2 + O_2^{*-} + H^{+}}, \qquad (4)$$

$$\mathrm{RH_2{+}O_2^{*-}{+}H \rightarrow RH^{*}{+}H_2O_2}, \qquad (5)$$

$$\mathrm{RH^{*}{+}O_2{+}H \rightarrow R + O_2^{*}{+}H^{+}}. \qquad (6)$$

Transition metals could be involved in the initiation of PPs redox cycling (Equations 4–6) as catalysts.[30,31]

Semiquinones (RH*), reactive oxygen species (ROS, O_2^{*-}, and H_2O_2), and the other short-lived by-products of PPs autoxidation may interact with biomolecules, initiating lipid peroxidation, protein and DNA oxidation, formation of DNA adducts, and endogenous antioxidant depletion.[35,36] It is generally accepted that the ability of PPs to undergo autoxidation and redox cycling is closely associated with their pro-oxidant action in biological systems,[37,38] which may lead to cytotoxicity,[39] mutagenicity, and carcinogenesis.[27]

PPs have been found to be strong chelators of metal ions, such as Fe^{2+}, Fe^{3+}, Cu^{2+}, Zn^{2+}, and Mn^{2+} (see Refs. 40 and 41). The redox properties of PP–metal complexes depend on the nature of the ligand, the strength of the metal–ligand bond, and the redox potential of the complex.[42,43] Both free radical–scavenging potency and cytoprotective activity of PP–metal complexes are sometimes significantly higher than those of the parent PPs.[44,45] At the

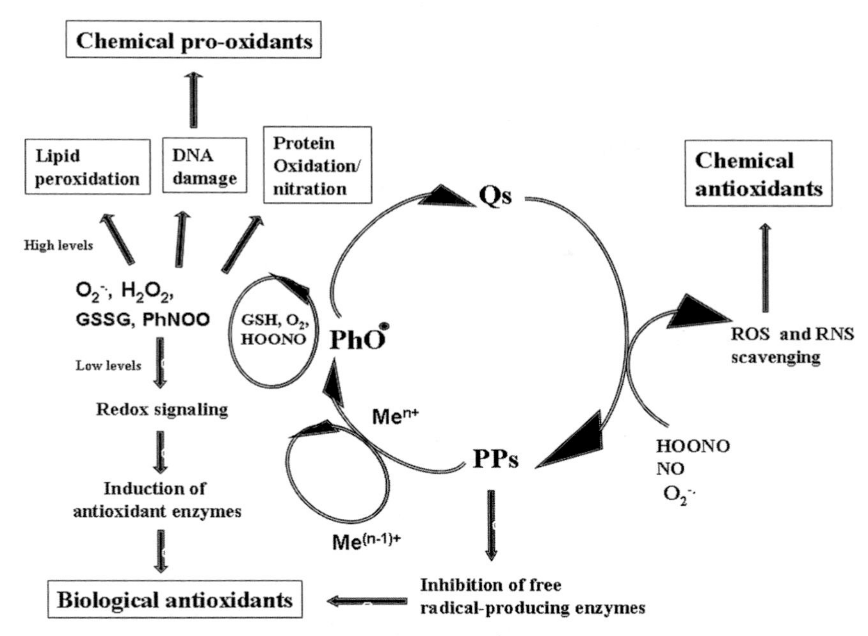

Figure 1. PPs as chemical and biological anti- and pro-oxidants. $PhO^{•}$, phenoxyl radicals; Qs, quinones; RNS, reactive nitrogen species; Me^{n+} and $Me^{(n-1)+}$, oxidized and reduced forms of transition metal ions; GSH and GSSG, reduced and oxidized forms of glutathione, respectively; PhNOO, nitrated PP.

same time, several PPs such as flavonoids, curcumin, resveratrol, and EGCG might mobilize copper from chromatin forming redox-active Cu–PPs complexes and hydroxyl radicals in a close vicinity to DNA, thus causing its cleavage.[46–48] Additionally, the PPs ligand in the PP–metal complexes could be easily oxidized, producing ROS and redox-dependent DNA strand breaks.[40,49,50]

Redox homeostasis in the cells and in the extracellular space may be seriously affected by PPs due to their interaction with the ROS-producing enzymes (cyclooxygenases, lipoxygenases, peroxidases, nitric oxide synthases, NADPH oxidases, and xanthine oxidase). PPs are known to challenge the enzymes: due to chemical similarity to their endogenous substrates,[22] by being their competitive inhibitors, by enhancing their enzymatic activity,[51] or/and by affecting the expression of corresponding genes.[1]

There is growing evidence that PPs induce an array of antioxidant (superoxide dismutase, glutathione peroxidase, gamma-glutamylcysteine synthase) and phase II detoxifying enzymes (glutathione-S-transferase, hemoxygenase-1, NADPH quinone oxidoreductase), by activating the transcription factor Nrf2, which upregulates genes containing antioxidant response element.[52,53] The Nrf2-connected mechanism is currently considered a major cause of indirect antioxidant action of PPs in biological systems.[24,54]

Collectively, depending on the chemical structure, redox potential, and the environment, PPs may possess direct chemical or indirect biological antioxidant properties as well as free-radical–generating/pro-oxidant activity (Fig. 1).

PP metabolism in the skin

The topically applied dermatological drugs, skin care products, cosmetics, aromatic oils, and perfumes, as well as environmental contact with plants or toxins, may be activated or inactivated by drug-metabolizing enzymes in the skin. As a consequence, activated metabolites may bring locally either health effects (anti-inflammatory, wound healing, angiogenic, chemotherapeutic, or antimicrobial ones) or undesirable adverse effects (skin sensitization, skin carcinogenesis, or phototoxic reactions in the skin). Keratinocytes contain many phase I (reduction/oxidation) and phase II (biotransformation) enzymes as compared to other skin cells—that is, fibroblasts, dendritic cells, Langerhans cells, leukocytes, etc.[55] Many exogenous chemical compounds, including PPs, are either substrates or inducers or inhibitors of the cytochrome P450 (CYP) superfamily of phase I enzymes.[56,57] UV irradiation is known to upregulate CYP1A1 and CYP1B1 in human keratinocytes[58,59] through activation of the aryl hydrocarbon receptor (AhR) pathway,[60] which could enhance bioactivation of environmental toxins (polycyclic aromatic hydrocarbons) or PPs, thus increasing substantially the risk of adverse phytophotoreactions in human skin such as contact dermatitis and carcinogenesis.[58] As aromatic hydrocarbons, PPs might affect the aryl hydrocarbon receptor (AhR), the transcription factor sensor for organic chemicals, which starts transcription of numerous detoxification genes coding for phase I and II metabolizing enzymes, particularly the cytochrome P450 CYP1 subfamily, Nrf2, and glutathione S-transferase (GST).[22,61,62] Another member of the CYP superfamily present in the skin, CYP3A4, is known to metabolize about 60% of clinically prescribed drugs and is regulated by the pregnane X receptor, a transcriptional factor that is activated by structurally diverse lipophilic xenobiotics, for example, dexamethasone.[55,61] Modulation of pregnane X receptor–mediated gene expression of CYP3A4 by single polyphenols (quercetin, curcumin, EGCG, and resveratrol) and polyphenol-rich extracts of allspice, clove, and thyme has been reported.[24]

Interaction of PPs with phase II detoxification enzymes has attracted much attention in the last few years.[57,61–63] In the skin, expression of phase II enzymes—glutathione-S-transferases (GST), UDP-glucoronosyl transferase (UGT), and catechol-*O*-methyl transferase (COMT)—have been found, and all of them are affected by polyphenols. The genetic element essential for mediating the induction has been identified as the antioxidant-responsive or electrophile-responsive elements (ARE or EpRE, respectively), which are present in the promoter regions of many genes encoding phase II enzymes. GST and other phase II detoxifying enzymes are regulated by the Nrf2/Keap 1 transcription factor pathway.[64] The functional requirements for Nrf2 activation include electrophilic properties, reactivity with thiols, and metal chelation.[24] PPs are good candidates to induce phase II enzymes through the Nrf2/Keap 1 system as they fulfill all of these requirements.

It has been hypothesized that peroxidases, such as myeloperoxidase, lactoperoxidase, thyroid peroxidase, and prostaglandin H synthase, could represent an alternative polyphenol-metabolizing pathway competing with CYPs in the skin.[65] The broad substrate specificities of these enzymes make them excellent candidates to catalyze the oxidation of PPs. The primary products of peroxidase-catalyzed oxidation of flavonoids are semiquinone or phenoxyl radicals, which having diffused from the active site of the enzyme can then oxidize molecules with low-redox potentials (GSH or NADH). This mechanism may be involved in the enzyme inactivation,[66] causing haptenization of endogenous proteins, which in turn may lead to a complete allergen formation. When reacting with tyrosine residues, phenoxyl radicals induce oxidative crosslinking of proteins, thus affecting their structural, enzymatic, and receptor properties.[67]

Being metabolic derivatives of amino acid phenylalanine, PPs share some structural and functional similarities with biologically active mammal molecules synthesized from phenylalanine as a precursor, such as catecholamines, thyroid and estrogen hormones, and melanin moieties.[68] Hence, PPs could easily interfere with metabolic processes and molecular pathways in the skin, bringing either desired health or adverse toxic effects. Interactions of PPs with metabolic systems of human skin and biological consequences of them are collected in Figure 2.

PPs and skin inflammation

An acute inflammatory response in the skin is a local protective reaction to various stresses including injury, microbial invasion, solar irradiation, and environmental pollutants. There is a steadily growing interest in skin protection from excessive inflammatory insults by PPs. Mechanistic studies have repeatedly shown that PPs with free-radical scavenging, antioxidant, and transition metal–chelating activity may directly diminish levels of non-protein inflammatory mediators, such as free radicals, hydrogen peroxide, lipid peroxides, and aldehydes, and final products of lipid peroxidation such as malonyl dialdehyde, 4-hydroxy-2-nonenal, and acrolein.[7,42] However, in the last 5–10 years, evidence from numerous *in vitro* skin cell studies suggests that PPs can influence cellular functions by multiple other mechanisms, such as direct interaction with several receptors, modulation of intracellular signal transduction and transcription of a number of genes,

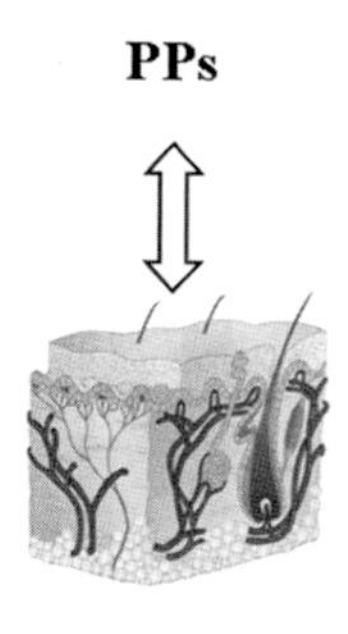

Figure 2. Metabolism and biological consequences of PPs in human skin. Numerous phase I (oxidation/reduction) and phase II (conjugation) enzymes are constitutively expressed, mainly in keratinocytes, and could be induced or suppressed by organic toxins, drugs, PPs, and UV irradiation. Biological consequences of such an interaction are diverse. CYP, cytochrome P450; MPO, myeloperoxidase; LPO, lactate peroxidase; EPO, eosinophil peroxidase; Gpx, glutathione peroxidase; HO, hemoxygenase; FMO, flavin monoxygenase; PGHS, prostaglandin H synthase; COX, cyclooxygenase; ADH, alcohol dehydrogenase; NADPH:QR, NADPH quinine reductase; GST, glutathione-*S*-transferase; UGT, UDP-glucoronyl transferase; COMT, catechol-*O*-methyl transferase.

post-translational modulation of enzymatic activities,[1,22] as well as epigenetic regulation of gene expression.

As documented by a vast literature, PPs inhibit major proinflammatory enzymes, such as inducible nitric oxide synthase (iNOS), NADPH oxidase, and eicosanoid-generating enzymes, such as cyclooxygenase (COX) and lipoxygenase (LOX), but also phospholipase A_2, which liberates arachidonic acid from membrane-bound phospholipids.[7,69–71] Hence, inhibition of this last enzyme prevents generation of arachidonic acid–derived mediators of inflammation. Furthermore, interaction of PPs with distinct receptors, such as peroxisome proliferator–activated receptors (PPARs) and estrogen receptors (ERs), results in the inhibition of inflammatory responses[69,72,73] through suppression of inflammatory gene transcription.[73–75] Distinct PPs, such as resveratrol, capsaicin, curcumin, epigallocatechin galate, genistein, and others,[69] are inducers of non-steroidal anti-inflammatory drug-activated gene-1, likewise, non-steroidal anti-inflammatory drugs.

Numerous experimental data confirm that PPs exert their anti-inflammatory action by modulating cellular inflammatory response regulated mainly by nuclear factor kappa B factor (NF-κB).[75,76] As a consequence, the NF-κB–dependent expression of proinflammatory proteins–cytokines, such as IL-1, IL-6, GM-CSFs, and TNF-α, is dramatically impaired or totally blocked.[77] Substantial experimental evidence points out that PPs—for example, apigenin and curcumin—act as potent inhibitors of activator protein (AP-1), a transcription factor controlling stress responses in a variety of human cells.[76,78] Moreover, PPs-induced abrogation of both factors NF-κB and AP-1 provides the best condition to oppose expression of proinflammatory mediators and may have a great impact on their anti-inflammatory action.[79]

Many regulatory elements, presumably implicated in the inflammatory cell responses, such as the antioxidant response element (ARE), protein kinase B (Akt), and extracellular signal-regulated protein kinase1/2 (ERK1/2),[80–82] are involved in

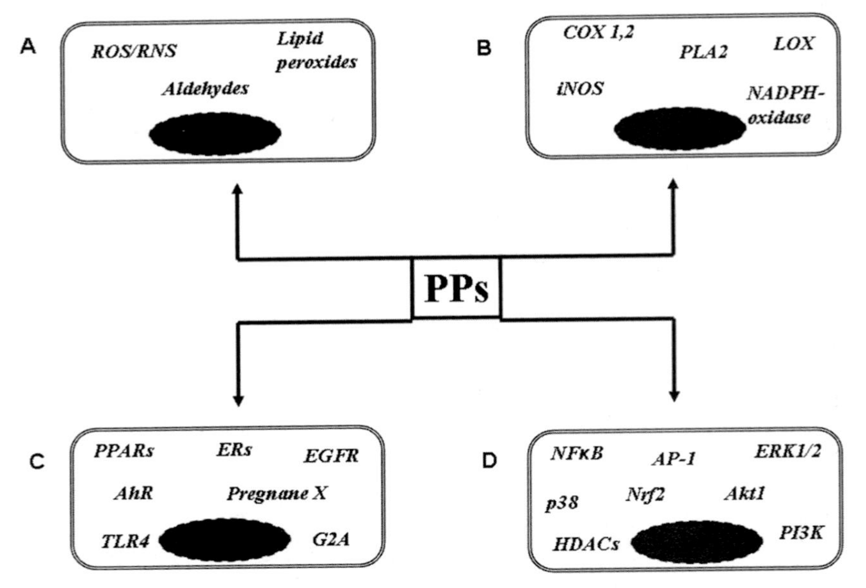

Figure 3. Multiple mechanisms of PP effects on inflammatory responses of human keratinocytes. (A) Interaction with non-protein inflammatory mediators: ROS/RNS, lipid peroxides, and aldehydes. (B) Interaction with proinflammatory enzymes: COX1 and COX2, iNOS, LOX, and NADPH oxidase. (C) Interaction with keratinocyte receptors involved into inflammatory response: PPARs; ERs; EGFR, epidermal growth factor receptor; xenobiotic receptors (AhR–aryl hydrocarbon and pregnane X); Toll-like receptor 4 (TLR4), and oxidized unsaturated fatty acid receptor (G2A). (D) Interference with transcription factors and epigenetic mechanisms involved in the inflammatory response: HDACs, histone deacetylases; PI3K, phosphoinositol-3-kinase; MAPK kinases, (ERK1/2 and p38); protein kinase B (Akt1); activator protein (AP-1).

signal transduction from PPs to PP-responsive genes. For comprehensive reviews on the subject, see Ref. 83.

Pro- and anti-inflammatory effects of PPs also strongly depend on the nature of the inflammatory response trigger used and the inflammatory marker evaluated. PP-modulated inflammatory responses in keratinocytes operated through the classical NF-κB–inhibiting mechanism and also targeted cytoplasmic and nuclear components of epidermal growth factor receptor (EGFR), EGFR-ERK axis, and AhR-machinery.[60,84] Importantly, AhR is interconnected with multiple signal transduction pathways, so that AhR could reasonably contribute to control major players of inflammatory responses in keratinocytes. Nuclear factor Nrf2 and related proteins belong to a family of xenobiotic sensors activated by oxidants and electrophiles, PPs among them. Effects of PPs on Nrf2 may be exerted either via redox control or through the AhR machine.[61,62] According to recent findings, the AhR pathway controls also expression of a number of proinflammatory factors, such as cytokines and growth factors, and downregulates NF-κB activation, which is a major player in cellular inflammatory responses.[84,85] In Figure 3, multiple mechanisms underlying anti- and proinflammatory effects of PPs are schematically shown.

Plant PPs are considered by botanical dermatologists as major contributors to the most common adverse cutaneous reactions, namely allergic and/or irritant contact dermatitis, which occur due to casual contact with a plant or occupational necessity to deal with plant material[19] or intentional application of herbal remedies onto the human skin.[20] For example, medicinal herbs that are known for causing adverse skin reaction (aloe, arnica, pineapple, calendula, chamomile, goldenseal, tea tree oil, and yarrow) also contain large amounts of PPs.[21] To exert allergenic or skin irritating properties, low-molecular weight phenolic molecules should possess either high hydrophobicity or contain a long hydrocarbon side chain. Bioactivation of a parent molecule (hapten) to reactive metabolite(s), followed by protein haptenation, has long been suggested as a critical step in the initiation of adverse cutaneous reactions to exogenous agents.[86] The protein modification by haptens occurs either by classical nucleophilic/electrophilic or by free radical-driven reactions.[87,88]

PP effects on UV–skin interaction

The damaging effects of UV light toward plant and mammal cells occur either by direct attack of biologically important molecules, such as DNA (Type I reaction), or *via* the generation of ROS, such as singlet oxygen and superoxide anion-radical (Type II reaction). In general, UV light is primarily absorbed by endogenous chromophores (trans-urocanic acid, melanins, porphyrins, flavins, quinones, chlorophil, plant pigments, tryptophan, hydropiridines, bilirubin, and glycation end-products).[43] Of note, the majority of PPs belong to plant pigments, hence, they are effective chromophores for UV light. After absorption of the light energy the chromophores become activated (excitation state). The activated chromophores may dissipate excessive energy in an exothermal process by emitting infrared radiation, thus playing an essential role as chemical UV filters. Alternatively, they directly react with a target molecule, thus transforming it into a corresponding free radical (type I photosensitization).[22] In the presence of oxygen, its reaction with an activated chromophore takes place, and superoxide radicals or singlet oxygen are formed. These ROS may induce oxidative modifications of lipids, proteins, and DNA (type II photosensitization). Collectively, PPs participate in photochemical reactions and, depending on their chemical structure and concentration, may diminish UV damage (photoprotective PPs) or aggravate it (phototoxic PPs). Quercetin and many other PPs protect human keratinocytes from UVA damage mainly by elevating intracellular antioxidant potential via the enhanced accumulation of Nrf2, a transcription factor for antioxidant genes.[89]

Although resveratrol is widely considered an effective photoprotective substance,[90] it potentiated the production of significant amounts of 8-oxo-7,8-dihydro-2′-deoxyguanosine in UVA-irradiated genomic DNA.[91] Owing to this observation, it is possible that topical application of resveratrol-containing products to skin exposed to sunlight could produce a potential hazardous action.

Conflicts of interest

The authors declare no conflicts of interest.

References

1. Korkina, L., V. Kostyuk, C. De Luca & S. Pastore. 2011. Plant polyphenols as emerging anti-inflammatory agents. *Mini Rev. Med. Chem.* **11:** 823–835.
2. Hsu, S. 2005. Green tea and the skin. *J. Am. Acad. Dermatol.* **52:** 1049–1059.
3. Yoon, J.H. & S.J. Baek. 2005. Molecular targets of dietary polyphenols with anti-inflammatory properties. *Yonsei Med. J.* **46:** 585–596.
4. Kim, H.P., K.H. Son, H.W. Chang & S.S.K. Kang. 2004. Anti-inflammatory plant flavonoids and cellular action mechanisms. *J. Pharmacol. Sci.* **96:** 229–245.
5. Lambert, J.D., J. Hong, G.Y. Yang, *et al.* 2005. Inhibition of carcinogenesis by polyphenols: evidence from laboratory investigations. *Am. J. Clin. Nutr.* **81:** 284S–91S.
6. Katiyar, S.K., M.S. Matsui & H. Mukhtar. 2000. Ultraviolet-B exposure of human skin induces cytochromes P450 1A1 and 1B1. *J. Invest. Dermatol.* **114:** 328–333.
7. Aggarwal, B.B. & S. Shishodia. 2006. Molecular targets of dietary agents for prevention and therapy of cancer. *Biochem Pharmacol.* **71:** 1397–1421.
8. Kundu, J.K., E.J. Chang, H. Fugii, *et al.* 2008. Oligonol inhibits UVB-induced COX-2 expression in HR-1 hairless mouse skin-AP-1 and C/EBP as potential upstream targets. *Photochem. Photobiol.* **84:** 399–406.
9. Korkina, L. 2007. Phenylpropanoids as naturally occurring antioxidants: from plant defense to human health. *Cell. Mol. Biol. (Noisy-le-Grand)* **53:** 15–25.
10. Tse, W.P., C.H.K. Cheng, C.T. Che, *et al.* 2007. Induction of apoptosis underlies the Radix Rubiae-mediated antiproliferative action of human epidermal keratinocytes: implications for psoriasis treatment. *Inter. J. Mol. Med.* **20:** 663–672.
11. Ahmad, N. & H. Mukhtar. 2004. Cytochrome P450: a target for drug development for skin diseases. *J. Invest. Dermatol.* **123:** 417–425.
12. Szczurko, O. & H.S. Boon. 2008. A systematic review of natural health product treatment for vitiligo. *BMC Dermatol.* **8**:2.
13. Wen-Jun, L., W. Hai-Yan, L. Wei, *et al.* 2008. Evidence that geniposide abrogates norepinephrin-induced hypopigmentation by the activation of GLP-1R-dependent c-kit signaling in melanocytes. *J. Ethnopharmacol.***118:** 154–158.
14. Korkina, L., E. Mikhal'chik, M. Suprun, *et al.* 2007. Molecular mechanisms underlying wound healing and anti-inflammatory properties of naturally occurring biotechnologically produced phenylpropanoid glycosides. *Cell. Mol. Biol. (Noisy-le-Grand)* **53:** 84–91.
15. Berger, M.M., M. Baines, W. Raffoul, *et al.* 2007. Trace element supplementation after major burns modulates antioxidant status and clinical course by way of increased tissue trace element concentrations. *Am. J. Clin. Nutr.* **85:** 1293–1300.
16. Krutmann, J. & D. Yarosh. 2006. Modern photoprotection of human skin. In *Skin Aging*. B. Gilchrest and J Krutmann, Eds.: 103–112. Springer. Berlin, Heidelberg, New York.
17. Stahl, W. & H. Sies. 2011. Photoprotection by dietary carotenoids: concept, mechanisms, evidence and future. *Mol. Nutr. Food Res.* doi: 10.1002/mnfr.201100232.

18. Katiyar, S.K. 2007. UV-induced immune suppression and photocarcinogenesis: chemoprevention by dietary botanical agents. *Cancer Lett.* **255:** 1–11.
19. Avalos, J. & H.I. Maibach. 2000. *Dermatologic Botany.* CRC Press. Boca Raton, London, New York, Washington, D.C.
20. Thornfeldt, C. 2005. Cosmeceuticals containing herbs: fact, fiction, and future. *Dermatol. Surg.* **31:** 873–880.
21. Bedi, M.K. & P.D. Shenefelt. 2002. Herbal therapy in dermatology. *Arch. Dermatol.* **138:** 232–242.
22. Korkina, L., S. Pastore, C. De Luca & V. Kostyuk. 2008. Metabolism of plant polyphenols in the skin: beneficial versus deleterious effects. *Curr. Drug Metab.* **9:** 710–729.
23. Lambert, J.D., S. Sang, & C.S. Yang. 2007. Possible controversy over dietary polyphenols: benefits vs. risks. *Chem. Res. Toxicol.*, **20:** 583–585.
24. Kluth, D., A. Banning, I. Paur, *et al.* 2007. Modulation of pregnane X receptor- and electrophile responsive element-mediated gene expression by dietary polyphenolic compounds. *Free Radic. Biol. Med.* **42:** 315–325.
25. Korkina, L. 2007. Phenylpropanoids as naturally occurring antioxidants: from plant defense to human health. *Cell. Mol. Biol.* **53:** 15–25.
26. Harborne, J.B. 1986. Nature, distribution, and function of plant flavonoids. In *Plant Flavonoids in Biology and Medicine.* Cody, Middleton & Harborne, Eds.: 15–24. Alan R. Liss. New York.
27. Skibola, C.F. & M.T. Smith. 2000. Potential health impacts of excessive flavonoid intake. *Free Radic. Biol. Med.* **29:** 375–383.
28. Fraga, C.G. 2007. Plant polyphenols: how to translate their *in vitro* antioxidant actions to *in vivo* conditions. *IUBMB Life* **59:** 308–315.
29. Osawa, T. 1999. Protective role of dietary polyphenols in oxidative stress. *Mech. Aging Dev.* **111:** 133–139.
30. Denisov, E. & I. Afanas'ev. 2005. *Oxidation and Antioxidants in Organic Chemistry and Biology*, CBC Taylor & Francis Group. Boca Raton, London, New York, Singapore.
31. Jovanovic, S.V., S. Steenken, M.G. Simic & Y. Hara. 1998. In: *Flavonoids in Health and Disease*, Rice-Evans & Packer, Eds.: 137–161. Marcel Dekker, Inc. New York.
32. Bors, W., C. Michel, & S. Schicora. 1995. Interaction of flavonoids with ascorbate and determination of their univalent redox potentials: a pulse radiolysis study. *Free Radic. Biol. Med.* **19:** 45–52.
33. Afanas'ev, I.B. 2007. Signaling functions of free radicals superoxide & nitric oxide under physiological & pathological conditions. *Mol. Biotechnol.* **37:** 2–4.
34. Hensley, K., K.S. Williamson, & R.A. Floyd. 2000. Measurement of 3-nitrotyrosine and 5-nitro-gamma-tocopherol by high-performance liquid chromatography with electrochemical detection. *Free Radic. Biol. Med.*, **28:** 520–528.
35. Sakihama, Y., M.F. Cohen, S.C. Grace, & H. Yamasaki. 2002. Plant phenolic antioxidant and pro-oxidant activities: phenolics-induced oxidative damage mediated by metals in plants. *Toxicology* **177:** 67–80.
36. Kagan, V.E. & Y.Y. Tyurina. 1998. Recycling and redox cycling of phenolic antioxidants. *Ann. N. Y. Acad. Sci.* **854:** 425–434.
37. Laughton, M.J., B. Halliwel, P.J. Evans, & J.R. Hoult. 1989. Antioxidant and pro-oxidant actions of the plant phenolics quercetin, gossypol, and myricetin. Effects on lipid peroxidation, hydroxyl radical generation and bleomycin-dependent damage to DNA. *Biochem. Pharmacol.* **38:** 2859–2865.
38. Canada, A.T., E. Giannella, T.D. Nguyen & R.P. Mason. 1990. The production of reactive oxygen species by dietary flavonols. *Free Radic. Biol. Med.* **9:** 441–449.
39. Canada, A.T., W.D. Watkins & T.D. Nguyen. 1989. The toxicity of flavonoids to guinea pig enterocytes. *Toxicol. Appl. Pharmacol.* **99:** 357–361.
40. Brown, J.E., H. Khodr, R. Hider, & C.A. Rice-Evans. 1998. Structural dependence of flavonoid interactions with Cu^{2+} ions: implications for their antioxidant properties. *Biochem. J.* **330:** 1173–1178.
41. Afanas'ev, I.B., A.I. Dorozhko, A.V. Brodskii, *et al.* 1989. Chelating and free radical scavenging mechanisms of inhibitory action of rutin and quercetin in lipid peroxidation. *Biochem. Pharmacol.* **38:** 1763–1769.
42. Halliwell, B. & J.M.C. Gutteridge. 2007. *Free Radicals in Biology and Medicine*, 4th ed. Oxford University Press. USA.
43. Galey, J.B. 1997. Potential use of iron chelators against oxidative damage. *Adv. Pharmacol.* **38:** 167–203.
44. Kostyuk, V.A., A.I. Potapovich, E.N. Strigunova, *et al.* 2004. Experimental evidence that flavonoids metal complexes may act as mimics of superoxide dismutase. *Arch. Biochem. Biophys.* **428:** 204–208.
45. Moridani, M.Y., J. Pourahmad, H. Bui, *et al.* 2003. Dietary flavonoid iron complexes as cytoprotective superoxide radical scavengers. *Free Radic. Biol. Med.* **34:** 243–253.
46. Hadi, S.M., H.B. Showket, S.A. Azmi, *et al.* 2007. Oxidative breakage of cellular DNA by plant polyphenols: a putative mechanism for anticancer properties. *Cancer Biol.* **17:** 370–376.
47. Azmi, A.S., S.H. Bhat, S. Hanif & S.M. Hadi. 2006. Plant polyphenols mobilize endogenous copper in human peripheral lymphocytes leading to oxidative DNA breakage: a putative mechanism for anticancer properties. *FEBS Lett.* **580:** 533–538.
48. Zheng, L.F., Q.Y. Wei, Y.J. Cai, *et al.* 2006. DNA damage induced by resveratrol and its synthetic analogues in the presence of Cu (II) ions: mechanism and structure-activity relationship. *Free Radic. Biol. Med.* **41:** 1807–1816.
49. Schmalhausen, E.V., E.B. Zhlobek, I.N. Shalova, *et al.* 2007. Antioxidant and pro-oxidant effects of quercetin on glyceraldehyde-3-phosphate dehydrogenase. *Food. Chem. Toxicol.* **45:** 1988–1993.
50. Jun, T., W. Bochu & Z. Liancai. 2007. Hydrolytic cleavage of DNA by quercetin manganese (II) complexes. Colloids Surf. B. *Biointerfaces* **55:** 149–152.
51. Borbulecych, O.Y., J. Jankun, S. H. Selman & E. Skrzypczak-Jankun. 2004. Lipoxygenase interactions with natural flavonoid, quercetin, reveal a complex with protocatechuic acid in its X-ray structure at 2.1 Å resolution. *Proteins* **54:** 13–19.
52. Zhang, D.D. 2006. Mechanistic studies of the Nrf2-Keap 1 signaling pathway. *Drug Metab. Rev.* **38:** 769–789.

53. Kobayashi, M. & M. Yamamoto. 2006. Nrf2-Keap 1 regulation of cellular defense mechanisms against electrophils and reactive oxygen species. *Adv. Enzyme Regul.* **46:** 113–140.
54. Chun, K.S. & Y.J. Surh. 2004. Signal transduction pathways regulating cyclooxygenase-2 expression: potential molecular targets for chemoprevention. *Biochem Pharmacol.* **68:** 1089–1100.
55. Baron, J.M., D. Holler, R. Schiffer, *et al.* 2001. Expression of multiple cytochrome p450 enzymes and multidrug resistance-associated transport proteins in human skin keratinocytes. *J. Invest. Dermatol.* **116:** 541–548.
56. Ahmad, N. & H. Mukhtar. 2004. Cytochrome p450: a target for drug development for skin diseases. *J. Invest. Dermatol.* **123:** 417–425.
57. Oesch, F., E. Fabian, B. Oesch-Bartlomowicz, *et al.* 2007. Drug-metabolizing enzymes in the skin of man, rat, and pig. *Drug Metab. Rev.* **39:** 659–698.
58. Katiyar, S.K., M.S. Matsui & H. Mukhtar. 2000. Ultraviolet-B exposure of human skin induces cytochromes P450 1A1 and 1B1. *J. Invest. Dermatol.* **114:** 328–333.
59. Villard, P.H., E. Sampol, J.L. Elkaim, *et al.* 2002. Increase of CYP1B1 transcription in human keratinocytes and HaCaT cells after UV-B exposure. *Toxicol. Appl. Pharmacol.* **178:** 137–143.
60. Potapovich, A.I., D. Lulli, P. Fidanza, *et al.* 2011. Plant polyphenols differentially modulate inflammatory responses of human keratinocytes by interfering with activation of transcription factors NFκB and AhR and EGFR-ERK pathway. *Toxicol. Appl. Pharmacol.* **255**:138–149.
61. Pavek, P. & Z. Dvorak. 2008. Xenobiotic-induced transcriptional regulation of xenobiotic metabolizing enzymes of the cytochrome P450 superfamily in human extrahepatic tissues. *Curr. Drug Metab.* **9:** 129–143.
62. Niestroy, J., A. Barabara, K. Herbst, *et al.* 2011. Single and concerted effects of benzo[a]pyrene and flavonoids on the AhR and Nrf2-pathway in the human colon carcinoma cell line Caco-2. *Toxicol. In Vitro* **25:** 671–683.
63. Lambert, J.D., S. Sang, A.Y. Lu & C.S.Yang. 2007. Metabolism of dietary polyphenols and possible interactions with drugs. *Curr. Drug. Metab.* **8:** 499–507.
64. Cermak, R. 2008. Effect of dietary flavonoids on pathways involved in drug metabolism. *Expert Opin. Drug Metab. Toxicol.* **4:** 17–35.
65. Itoh, K., T. Chiba, S. Takahashi, *et al.* 1997. An Nrf2/small Maf heterodimer mediates the induction of phase II detoxifying enzyme genes through antioxidant response elements. *Biochem. Biophys. Res. Commun.* **236:** 313–322.
66. Strohm, B.H. & A.P. Kulkarni. 1986. Peroxidase, an alternate pathway to cytochrome P-450 for xenobiotic metabolism in skin: partial purification and properties of the enzyme from neonatal rat skin. *J. Biochem. Toxicol.* **1:** 83–97.
67. Divi, R.L. & D.R. Doerge. 1994. Mechanism-based inactivation of lactoperoxidase and thyroid peroxidase by resorcinol derivatives. *Biochem.* **33:** 9668–9674.
68. Heinecke, J.W., W. Li, G.A. Francis & J.A. Goldstein. 1993. Tyrosyl radical generated by myeloperoxidase catalyzes the oxidative crosslinking of proteins. *J. Clin. Invest.* **91:** 2866–2872.
69. Kim, Y.J., J.K. No, J.H. Lee, & H.Y. Chung. 2005. 4,4-Dihydroxybiphenyl as a new potent tyrosinase inhibitor. *Biol. Pharm. Bull.*, **28:** 323–327.
70. Ramos S. 2008. Cancer chemoprevention and chemotherapy: dietary polyphenols and signalling pathways. *Mol. Nutr. Food Res.* **52:** 507–526.
71. Esposito, E., E. Mazzon, I. Paterniti, *et al.* 2010. PPAR-alpha contributes to the anti-inflammatory activity of verbascoside in a model of inflammatory bowel disease in mice. *PPAR Res.* 917312.
72. Simonyi, A., D. Woods, A.Y. Sun & G.Y. Sun. 2002. Grape polyphenols inhibit chronic ethanol-induced COX-2 mRNA expression in rat brain. *Alcohol Clin. Exp. Res.* **26:** 352–357.
73. Lee, J.L., H. Mukhtar, D.R. Bickers, *et al.* 2003. Cyclooxygenases in the skin: pharmacological and toxicological implications. *Toxicol. Appl. Pharmacol.* **192:** 294–306.
74. Jiang, C., A.T. Ting & B. Seed. 1998. PPAR-gamma agonists inhibit production of monocyte inflammatory cytokines. *Nature* **391:** 82–86.
75. Kim, S., H.J. Shin, S.Y. Kim, *et al.* 2004. Genistein enhances expression of genes involved in fatty acid catabolism through activation of PPARalpha. *Mol. Cell. Endocrinol.* **220:** 51–58.
76. Klaunig, J.E., M.A. Babich, K.P. Baetcke, *et al.* 2003. PPAR-alpha agonist-induced rodent tumors: modes of action and human relevance. *Crit. Rev. Toxicol.* **33:** 655–780.
77. Kawanishi, S. & Y. Hiraku. 2001. Sequence-specific DNA damage induced by UVA radiation in the presence of endogenous and exogenous photosensitizers. In: *Oxidants and Antioxidants in Cutaneous Biology*. J. Thiele & P. Elsner, Eds.: 74–82. S. Karger AG. Basel, Switzerland.
78. Rahman, I., S.K. Biswas & P.A. Kirkham. 2006. Regulation of inflammation and redox signaling by dietary polyphenols. *Biochem. Pharmacol.* **72:** 1439–1452.
79. Portugal, M., V. Barak, I. Ginsburg & R. Kohen. 2007. Interplay among oxidants, antioxidants, and cytokines in skin disorders: present status and future considerations. *Biomed. Pharmacother.* **61:** 412–422.
80. Balasubramanian, S. & R.L. Eckert. 2007. Keratinocyte proliferation, differentiation, and apoptosis– differential mechanisms of regulation by curcumin, EGCG and apigenin. *Toxicol. Appl. Pharmacol.* **224:** 214–219.
81. Chen, C., R. Yu, E.D. Owuor & A.N. Kong. 2000. Activation of antioxidant-response element (ARE), mitogen-activated protein kinases (MAPKs) and caspases by major green tea polyphenol components during cell survival and death. *Arch. Pharm Res.* **23:** 605–612.
82. Na, H.K., E.H. Kim, J.H. Jung, *et al.* 2008. (-)-Epigallocatechin gallate induces Nrf2-mediated antioxidant enzyme expression via activation of PI3K and ERK in human mammary epithelial cells. *Arch. Biochem. Biophys.* **476:** 171–177.
83. Siow, R.C.M., F.Y.L. Li, D.J. Rowlands, *et al.* 2007. Cardiovascular targets for estrogens and phytoestrogens: transcriptional regulation of nitric oxide synthase and antioxidant defense genes. *Free Radic. Biol. Med.* 42: 909–925.
84. Pastore, S., D. Lulli, P. Fidanza, *et al.* 2012. Plant polyphenols regulate chemokine expression and tissue repair in human keratinocytes through interaction with cytoplasmic and

nuclear components of epidermal growth factor receptor system. *Antioxid. Redox Signal.* **16:** 314–328.
85. Haarmann-Stemmann, T., H. Both & J. Abel. 2009. Growth factors, cytokines and their receptors as downstream targets of arylhydrocarbon receptor (AhR) signaling pathways. *Biochem. Pharmacol.* **77:** 508–520.
86. Roychowdhury, S., A.E. Cram, A. Aly, & C.K. Svensson. 2007. Detection of haptenated proteins in organotypic human skin explant cultures exposed to dapsone. *Drug Metab. Dispos.* **35:** 1463–1465.
87. Karlberg, A.T., M.A. Bergström, A. Börje, *et al.* 2008. Allergic contact dermatitis–formation, structural requirements, and reactivity of skin sensitizers. *Chem. Res. Toxicol.* **21:** 53–69.
88. Schmidt, R.J., L. Khan & L.Y. Chung, 1990. Are free radicals and not quinones the haptenic species derived from urushiols and other contact allergenic mono- and dihydric alkylbenzenes? The significance of NADH, glutathione, and redox cycling in the skin. *Arch. Dermatol. Res.* **282:** 59–64.
89. Kimura, S., E. Warabi, T. Yanagawa, *et al.* 2009. Essential role of Nrf2 in keratinocyte protection from UVA by quercetin. *Biochem. Biophys. Res. Commun.* **387:** 109–114.
90. Nichols, J.A. & S.K. Katiyar. 2010. Skin photoprotection by natural polyphenols: anti-inflammatory, antioxidant and DNA repair mechanisms. *Arch. Dermatol. Res.* **302:** 71–83.
91. Seve, M., F. Chimienti, S. Devergnas, *et al.* 2005. Resveratrol enhances UVA-induced DNA damage in HaCaT human keratinocytes. *Med. Chem.* **1:** 629–633.

Ann. N.Y. Acad. Sci. ISSN 0077-8923

ANNALS OF THE NEW YORK ACADEMY OF SCIENCES
Issue: *Environmental Stressors in Biology and Medicine*

Flavonoids and metabolic syndrome

Monica Galleano,[1] Valeria Calabro,[1] Paula D. Prince,[1] María C. Litterio,[1] Barbara Piotrkowski,[1] Marcela A. Vazquez-Prieto,[2] Roberto M. Miatello,[2] Patricia I. Oteiza,[3,4] and Cesar G. Fraga[1,3]

[1]Physical Chemistry, School of Pharmacy and Biochemistry, University of Buenos Aires, and Institute of Molecular Biochemistry and Medicine (IBIMOL)-CONICET, Buenos Aires, Argentina. [2]Department of Pathology, School of Medicine, National University of Cuyo and Laboratory of Cardiovascular Pathophysiology, Institute for Experimental Medical and Biological Research (IMBECU)-CONICET, Mendoza, Argentina. [3]Department of Nutrition, University of California, Davis, California. [4]Department of Environmental Toxicology, University of California, Davis, California

Address for correspondence: Monica Galleano, Physical Chemistry, School of Pharmacy and Biochemistry, UBA. Junín 956, C1113AAD, Buenos Aires, Argentina. mgallean@ffyb.uba.ar

Increasing evidence indicates that several mechanisms, associated or not with antioxidant actions, are involved in the effects of flavonoids on health. Flavonoid-rich beverages, foods, and extracts, as well as pure flavonoids are studied for the prevention and/or amelioration of metabolic syndrome (MS) and MS-associated diseases. We summarize evidence linking flavonoid consumption with the risk factors defining MS: obesity, hypertriglyceridemia, hypercholesterolemia, hypertension, and insulin resistance. Nevertheless, a number of molecular mechanisms have been identified; the effects of flavonoids modifying major endpoints of MS are still inconclusive. These difficulties are explained by the complex relationships among the risk factors defining MS, the multiple biological targets controlling these risk factors, and the high number of flavonoids (including their metabolites) present in the diet and potentially responsible for the *in vivo* effects. Consequently, extensive basic and clinical research is warranted to assess the final relevance of flavonoids for MS.

Keywords: polyphenols; obesity; inflammation; lipids; hypertension; diabetes

Metabolic syndrome, diet, and flavonoids

Metabolic syndrome and diet

In 1988, Reaven first described syndrome X,[1] which later was renamed metabolic syndrome (MS). In 2005, a consensus criterion was attained for the diagnosis of MS by incorporating definitions from both the International Diabetes Federation and the American Heart Association/National Heart, Lung, and Blood Institute. MS is diagnosed when three out of the five following risk factors are present: elevated waist circumference (according to population- and country-specific definitions); blood triglycerides (TG) $\geq$ 150 mg/dL; blood high density lipoproteins (HDL)-cholesterol $\leq$ 40 or 50 mg/dL in men and women, respectively; blood pressure (BP) $\geq$ 130/85 mmHg; and fasting glucose $\geq$ 100 mg/dL.[2]

The relationship between MS and diet was recently analyzed in two epidemiological studies: a cross-sectional study showing that the consumption of a diet consistent with the recommendations in the *Dietary Guidelines for Americans* is associated with lower values of several of the risk factors defining MS (i.e., waist circumference, BP, blood glucose, and blood triglycerides),[3] and a population-based study showing an association between a lower risk of MS and the adherence to the French nutritional guidelines (*Program National Nutrition Santé*).[4] Importantly, both American and French guidelines strongly recommend a high consumption of fruits and vegetables.[5,6] We propose that the presence of flavonoids in fruits and vegetables can in part explain their beneficial effects on MS.

Flavonoids and diet

Plants produce a large amount of polyphenolic compounds that act as secondary metabolites in plant physiology. According to their chemical structures, these polyphenols are divided into several families, one of which is the flavonoid family.[7] Edible plants,

doi: 10.1111/j.1749-6632.2012.06511.x

as well as foods and beverages derived from them, provide the human diet with generous amounts (up to 1 g/day) of flavonoids.[8] Flavonoids have a basic chemical structure constituted by two benzene rings linked through a heterocyclic pyran ring. Hydroxyl group substituents provide centers for reaction. Flavonoids can be divided into several subfamilies according to the degree of oxidation of the oxygen heterocycle and the substitution patterns, for example, flavones, flavonols, isoflavones, anthocyanins, flavanols, and flavanones.[9]

The understanding of the molecular mechanism by which a food component can mediate a physiological effect is necessary to design diets, to consider supplementation and/or to plan pharmacological approaches. During many years, the beneficial effects of flavonoids on health were ascribed to their antioxidant capacity that results from flavonoids' chemical possibilities of scavenging free radicals and/or chelate prooxidant metals. Increasing evidence suggests that a variety of mechanisms, associated or not with antioxidant effects, are involved in the actions of flavonoids in biological systems.[10–15] Based on these mechanisms, many flavonoid-rich beverages, foods, and extracts, as well as pure flavonoids are being studied as potential agents for the prevention and amelioration of MS-associated risk factors. This short review will focus on the flavonoids that are more abundant in the human diet, and/or on those that have been more studied in terms of modifying the risk factors defining MS. It should also be noted that the effects of flavonoids on the different risk factors defining MS will be addressed separately, although many of these factors can share a mechanism, for example, inflammation related to obesity, diabetes, and hypertension.

Metabolic syndrome factors and dietary flavonoids

Obesity and dietary flavonoids

Obesity is characterized by the accumulation of large amounts of fat in adipocytes, as well as by an increase in adipocyte size and number.[16] Solid evidence support a correlation between an excess of intra-abdominal or visceral adipose tissue (waist circumference) and other risk factors observed in MS.[17,18] In addition, excessive fat is responsible for the production of chemical mediators that relate obesity and inflammation.

A series of interventional studies were done in overweight and obese subjects to test the effects of different flavonoid-rich foods or beverages on obesity. The consumption of a polyphenol-rich cloudy apple juice (a mix of flavonoid and nonflavonoid compounds at 750 mL/day) for four weeks caused a significant reduction of the percentage of total body fat and an increment in lean body mass in obese subjects.[19] However, other studies showed no effects of flavonoids on obesity parameters. For example, in overweight/obese subjects, the consumption of dark chocolate rich in catechin and epicatechin (1.0 g/day) for two weeks[20] or administration of a licorice flavonoid oil (300 mg/day) for eight weeks[21] did not modify body mass, composition, or waist circumference. In another study of obese subjects flavonol quercetin supplementation (500 and 1000 mg/day) for 12 weeks did not modify body mass and body composition.[22] However, it should be noted that changes in body mass in obese/overweight individuals need strong, long-term interventions to result in a significant change in body weight or waist circumference. For example, the supplementation with a flavonoid at concentrations compatible with those present in a normal human meal could hardly produce a reduction of 5 kg in a person weighing 100 kg (5%), in a 15-day period.

Special attention has been paid to green tea consumption and obesity. A metaanalysis performed on 15 randomized controlled trials evaluated the effects of green tea catechins ((–)-epicatechin, (+)-catechin, (–)-epigallocatechin, (–)-epigallocatechin gallate (EGCG), and (–)-epicatechin gallate) on body mass index, body weight, and waist circumference, concluding that, only when combined with caffeine, these catechins reduced those parameters.[23] These associations could be the result of the effects of green tea catechins on energy expenditure and fat oxidation.[24] Additionally, green tea catechins could reduce glucose absorption by inhibiting gastrointestinal enzymes related to sugar metabolism like α-amylase and α-glucosidase.[25] Flavonoids have also been reported to decrease lipid absorption by inhibiting the activity of pancreatic lipase.[26,27] Flavonoids can directly act on the enzyme active site or, indirectly, by increasing lipid (triglyceride) droplet size, which reduces the substrate accessibility to the enzyme.[27,28]

Studies in various experimental models explored the actions of flavonoids on certain pathways of lipid metabolism. For example, a mulberry water extract containing flavonoid and nonflavonoid polyphenols administered to high fat-fed hamsters for 12 weeks promoted lipolysis and prevented hepatic lipogenesis.[29] This was linked to an increased hepatic expression of peroxisome proliferator-activated receptor (PPAR)α and carnitine palmitoyltransferase-1, which are involved in lipolysis, and to a decreased expression of enzymes involved in lipogenesis, including fatty acid synthase (FAS) and 3-hydroxy-3-methylglutaryl (HMG)-coenzyme A reductase.

When administered to obese-diabetic mice, tiliroside, a glycosidic flavonoid present in strawberries, increased whole-body fatty acid oxidation but failed to prevent body weight gain and visceral fat accumulation.[30] The increase in fatty acid oxidation was mediated by the activation of the adiponectin-mediated signaling through an upregulation of the mRNA expression of hepatic adiponectin receptors AdipoR-1 and AdipoR2, and skeletal muscle AdipoR1.

In summary, there is not enough evidence to attribute an effect of flavonoids on major end-points of obesity. However, the strong possibility that their beneficial actions may be due to their capacity to improve adipocyte functionality and fatty oxidation guarantees further investigations. It should be stressed that the flavonoids and flavonoid mixtures discussed above have different chemical structures and could thus exert their action through more than one mechanism.

Obesity, inflammation, and dietary flavonoids

The adipose tissue is an endocrine organ that secretes a variety of proinflammatory adipocytokines, including plasminogen activator inhibitor-1, tumor necrosis factor alpha (TNF-α), interleukin-6, monocyte chemoattractant protein-1, resistin, leptin, and adiponectin. Increased visceral adiposity is associated with a higher production of these adipocytokines leading to local and generalized inflammation.[31]

Among the adipocytokines, TNF-α is of particular relevance. TNF-α activates proinflammatory signaling cascades, such as the mitogen-activated protein kinases (MAPKs) and AP-1 (activator protein-1). The activation of these pathways downregulates PPARγ and induces the transcription of inflammatory genes that both maintain a sustained inflammatory state and impair insulin signaling, leading to obesity-triggered insulin resistance.[32]

Flavonoids and flavonoid-containing foods interfere with inflammatory signaling.[33] In fructose-fed rats, administration of flavonoid-containing red wine reduced adipose tissue weight and favored antiobesity (antiinflammation) pathways by reducing resistin expression (proobesity) and increasing adiponectin expression (antiobesity).[34] In obese Zucker rats, a 10-week administration of quercetin (10 mg/kg of body weight/day) increased plasma concentration of adiponectin, reduced TNF-α secretion and the expression of the proinflammatory iNOS (inducible nitric oxide synthase) in visceral adipose tissue.[35] In high fat–fed rodents, the administration of EGCG (3.2 g/kg diet to rats for 16 weeks[36] or 30 mg/kg grape-seed procyanidins to mice for 19 days[37]) reduced adipose tissue inflammation.

In a coculture of adipocytes and macrophages, 10–20 μg/mL of oligomerized grape seed polyphenols (essentially constituted by 4-*S*-cysteine derivatives of procyanidins B-1 and B-2) ameliorated inflammation by reducing NF-κB–regulated transcriptional activity, and downregulating phosphorylation of ERK1/2 (extracellular-regulated kinase 1/2).[38] In human macrophages and adipocytes treated with macrophage-conditioned media, quercetin (3–30 μM) attenuated the activation of NF-κB and MAPKs, decreasing not only parameters of inflammation but also of insulin resistance.[39]

In addition, EGCG (100 μM) reduced the expression of the proinflammatory adipocytokine resistin via in 3T3-L1 adipocytes.[40] (–)-Epicatechin was recently found to inhibit TNF-α–mediated NF-κB, and MAPKs activation and PPARγ downregulation in 3T3-L1 adipocytes.[41] These effects were associated with a decreased expression of proteins involved in inflammation and in insulin resistance.

Dyslipidemias and dietary flavonoids

Dyslipidemias are defined as abnormal amounts of lipids in the blood. MS dyslipidemia is characterized by elevated levels of TG, low-density lipoproteins (LDL) cholesterol, and low levels of HDL cholesterol. Flavonoid-rich foods or beverages and/or purified flavonoids, have been shown to lower plasma TG and/or total cholesterol and LDL cholesterol (or increase HDL cholesterol) in circulation in both humans with MS and rodent models of MS.[35,36,42–50] As previously discussed, flavonoids could be decreasing lipid absorption at the gastrointestinal level. However, flavonoids can also modulate the activity of different enzymes involved in lipid metabolism and the expression of transcription factors involved in TG and cholesterol synthesis, for example, sterol regulatory element-binding proteins SREBP-1, and SREBP-2.[51]

A grape-seed proanthocyanidin extract (25 mg/kg of body weight/day for 10 days) was found to reduce postprandial TG in normolipidemic rats[52] and mice,[53] and in high-fat fed rats,[54] in association with the downregulation of hepatic SREBP-1. Additionally, the same extract suppressed the over expression induced by a high-fat diet of enzymes involved in lipogenesis: FAS and ATP-citrate lyase isoform;[55] microsomal transfer protein, the key controller of very low-density lipoprotein assembly; and diglyceride acyltransferase 2.[54] Similarly, other flavonoid extracts, such as licorice, and purified flavonoids, such as baicalin, decreased liver SREBP-1, acetyl-CoA carboxylase, and FAS expression[56,57] and activity[58] in rats fed a high-fat diet. EGCG administration (3.2 g/kg diet, 16 weeks) to high fat-fed mice decreased body and liver weight and TG, and plasma cholesterol levels.[36]

In LDL receptor-null mice with diet-induced insulin resistance, the flavonone naringenin (3 g/100 g of the diet for four weeks) accelerated hepatic FA oxidation and prevented TG accumulation in the liver, effects associated with an increased expression of PPAR-γ coactivator-1α, carnitine acyltransferase I, and acyl-CoA oxidase, molecules involved in FA oxidation. Naringenin also reverts the increased expression of SREBP-1 in liver and muscle of rats, diminishing the SREBP-1c stimulated lipogenesis.[46] In a six-month treatment, narigerin lowered total hepatic cholesterol and cholesteryl ester.[49] Recently, the effects of administrating a cranberry extract enriched in flavonoids (2% of the diet for eight weeks) were evaluated in both nonobese and obese C57/BL6 mice. At the end of the treatment, nonobese mice showed a reduced expression of hepatic key enzymes involved in cholesterol synthesis: HMG coenzyme A-reductase, HMG coenzyme A-synthase, farnesyl diphosphate synthase, and squalene synthase. On the other hand, in obese mice, cranberry extract administration improved the lipid profile and reduced visceral fat mass. These effects are likely mediated by the activation of the adiponectin/AMPK (5′ AMP-activated protein kinase) pathway, which accelerates FA utilization in the muscle and decreases fat storage in the adipocytes.[48] Finally, in a type 2 diabetes model (db/db mice) the administration of a proanthocyanidin extract of persimmon peel (10 mg/kg body weight/day) caused the downregulation of SRBEP-1 and SREBP-2.[59]

In summary, flavonoids may improve dyslipidemias by modulating lipid absorption and lipogenesis.

Blood pressure and dietary flavonoids

The regulation of BP is the result of the interaction of a network of molecules and systems. We will focus on the bioavailability of NO (nitric oxide) that plays a central role in this complex scenario. Adequate NO levels are associated with appropriate vasodilation and normal BP, while decreases in NO bioavailability led to failure in smooth muscle relaxation resulting in BP increase. The maintenance of NO steady-state levels has multiple steps susceptible to be regulated by flavonoids. In fact, several flavonoids cause a BP lowering effect in normotensive and hypertensive subjects, (reviewed in Refs. 60 and 61).

In subjects with MS or presenting some of the risk factors associated with MS, supplementation with different flavonoids resulted in lowering BP. Berries were found effective in reducing systolic and diastolic BP (SBP and DBP) in patients with MS-associated risk factors (reviewed in Ref. 43). In overweight/obese subjects, a polyphenol-rich dark chocolate (0.5–1.0 g/day for two weeks) reduced SBP and DBP; [20] cocoa flavanols administered at 902 mg/day for 12 weeks lowered DBP and mean arterial BP in association with an increase in flow-mediated dilation.[62] By contrast, the administration of hesperidin (0.5 g/day for three weeks) did not reduce BP in MS patients but resulted in an improvement in flow-mediated dilation; [63] EGCG administration to subjects with MS did not affect BP.[64]

Flavonoids also showed BP-lowering effects in rat models of MS.[34,47,65–67] In many of these studies, improved endothelial function and increased NO bioavailability were associated with lower BP. Augmentation in NO bioavailability can be the result of higher NO synthesis resulting from increased

NOS expression and/or activity or lower superoxide anion-mediated NO degradation. NADPH oxidase (NOX) is considered the most important source of superoxide anion in the vascular wall, thus, lower expression and/or activity of NOX would be associated with decreased superoxide anion production and the consequent diminished NO consumption. The hypothesis that flavonoids act through a reduction of NO degradation is supported by numerous studies in animal models. In Otsuka Long-Evans Tokushima Fatty (OLETF) rats, the long-term administration of (+)-catechin, reduced the spontaneous increase in BP and prevented endothelial dysfunction, in association with a decrease in NOX activity and expression of its subunits, $p47^{phox}$ and $p22^{phox}$, in thoracic aorta.[66] In parallel, enhanced eNOS activity and decreased expression of NOX membrane subunit Nox1 were detected. In a fructose-fed rat model, the administration of red wine and dealcoholized red wine reduced NOX activity in aortic tissue[34,67] and NOX subunit Nox4 expression in mesenteric tissue.[34] Even in MS models where BP was not increased, such as in the Zucker fatty rat model, treatment with a red wine polyphenols extract corrected endothelial dysfunction.[68]

The effects of flavonoids on hypertension are very well documented and appear to be mechanistically related to NO bioavailability regulated through the activation of NOS and/or the inhibition of NOX.

Insulin resistance and dietary flavonoids

Insulin resistance leads to the impairment of glucose tolerance, hyperinsulinemia and plays an important role in the development of type 2 diabetes. A role for flavonoids preventing/reverting insulin resistance has been suggested by improvements in several of these parameters in humans and animal models supplemented with flavonoids.[20,30,35,36,44,46,69] A recent meta-analysis showed that short-term consumption of flavanol-rich cocoa was associated with an improvement in insulin sensitivity in humans.[70] By contrast, two prospective studies did not show evidence of an association between flavonol or flavone consumption and markers of insulin resistance and type 2 diabetes in women,[71] and in a cohort of male smokers.[72] These observations stress the relevance of a mechanistic approach to the problem, considering not only the different physiological aspects but the flavonoids ascribed to a particular effect.

Mechanistically, adipose tissue insulin resistance is associated with increased fat storage in adipocytes and the development of local inflammation. Beneficial effects of flavonoids on these aspects have been detailed in previous sections. Hepatic inflammation and oxidative stress, considered inductors of hepatic insulin resistance, were ameliorated by administration of a flavonoid to fructose-fed rats for 60 days. Galangin (100 μg/kg) prevented the rise in plasma glucose and insulin, oxidative damage and

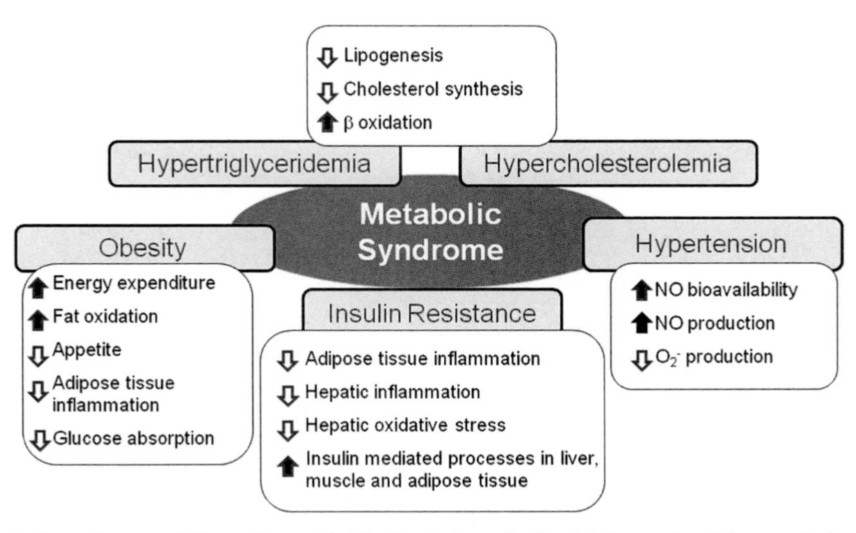

Figure 1. Metabolic syndrome and dietary flavonoids. The figure shows the five risk factors that define metabolic syndrome (gray boxes) and the potential beneficial effects of flavonoids on parameters associated with each factor (white boxes). The arrows indicate the direction of the changes mediated by flavonoids.

inflammatory changes in liver associated with fructose administration.[73]

The defects in insulin signaling leading to insulin resistance can occur at different levels: insulin binding to its receptor, insulin receptor autophosphorylation, tyrosine phosphorylation of insulin receptor substrates, recruitment and activation of phosphoinositide 3 (PI-3) kinase, production of phosphatidylinositol (3,4,5)-triphosphate, and/or activation of the protein kinase B pathway. In rats fed a high-fructose diet, green tea extract supplementation was evaluated in its ability to modulate the expression of genes involved in glucose transporters and the insulin-signaling pathway in liver and muscle. Results showed protective effects but the level of modulation was variable depending on the administered dose and the organ studied.[74] In the same line, green tea polyphenols (200 mg/kg body weight for six weeks) administered to fructose-fed rats, improve the mRNA and protein levels of insulin-signaling proteins (e.g., insulin receptor, insulin receptor substrate 1 and 2, PI-3 kinase, protein kinase B, glucose transporters 1 and 4) in cardiac muscle.[75] With the same fructose model, grape seed extract enhanced the expression of protein kinase B and glucose transporter 4 in skeletal muscle.[76] Kaempferitrin treatment (10–20 μM) of 3T3-L1 adipocytes activated the insulin transduction pathway by upregulating the phosphorylation of the insulin receptor β and insulin receptor substrate 1, and ser473 site in protein kinase B (PKB/Akt).[77] In summary, although limited, current evidence suggest that select flavonoids can regulate cellular events that lead to an improvement of insulin sensitivity. However, further investigations using isolated and pure flavonoids are necessary to provide support for such an antidiabetic function.

Conclusions

Flavonoids, as components of fruits and vegetables and the foods derived from them, can provide benefits for health. These benefits can be related to the participation of flavonoids in the control of the risk factors involved in MS (Fig. 1). To understand the impact of dietary flavonoids on MS implies the analysis of a very complex matrix: the risk factors defining the MS, the multiple biological targets involved in the onset of these risk factors, and the number of flavonoids (including their metabolites) that potentially are responsible for the *in vivo* actions. As a consequence, intense basic and applied (clinical) research is necessary to better understand the role of flavonoids in the prevention and/or amelioration of MS and MS-associated diseases.

Acknowledgments

This work was supported by grants of the University of Buenos Aires, UBACyT (20020090100111 and 20020100100659). MG, BP, MAV-P, RMM, PIO, and CGF are members of the Scientific Investigator Career of CONICET, Argentina. VC, PDP, and MCL hold CONICET fellowships.

Conflicts of interest

The authors declare no conflicts of interest.

References

1. Reavan, G.M. 1988. Banting lecture 1988. Role of insulin resistance in human disease. *Diabetes* **37:** 1595–1607.
2. Alberti, K.G. *et al.* 2009. Harmonizing the metabolic syndrome: a joint interim statement of the International Diabetes Federation Task Force on Epidemiology and Prevention; National Heart, Lung, and Blood Institute; American Heart Association; World Heart Federation; International Atherosclerosis Society; and International Association for the Study of Obesity. *Circulation* **120:** 1640–1645.
3. Fogli-Cawley, J.J. *et al.* 2007. The 2005 Dietary Guidelines for Americans and risk of the metabolic syndrome. *Am. J. Clin. Nutr.* **86:** 1193–1201.
4. Kesse-Guyot, E. *et al.* 2011. Adherence to French nutritional guidelines is associated with lower risk of metabolic syndrome. *J. Nutr.* **141:** 1134–1139.
5. USDA. 2005. *Dietary Guidelines for Americans.* US Department of Agriculture and Department of Health and Human Services 6th ed. Washington, DC.
6. Estaquio, C. *et al.* 2008. The French National Nutrition and Health Program score is associated with nutritional status and risk of major chronic diseases. *J. Nutr.* **138:** 946–953.
7. Manach, C. *et al.* 2004. Polyphenols: food sources and bioavailability. *Am. J. Clin. Nutr.* **79:** 727–747.
8. Scalbert, A. & G. Williamson. 2000. Dietary intake and bioavailability of polyphenols. *J. Nutr.* **130:** 2073S–2085S.
9. Jaganath, I.B. & A. Crozier. 2010. Dietary flavonoids and phenolic compounds. In *Plant Phenolics and Human Health: Biochemistry, Nutrition, and Pharmacology*. C.G. Fraga, Ed.: 1–4. John Wiley & Sons. Hoboken, New Jersey.
10. Frei, B. & J.V. Higdon. 2003. Antioxidant activity of tea polyphenols *in vivo*: evidence from animal studies. *J. Nutr.* **133:** 3275S–3284S.
11. Halliwell, B. *et al.* 2005. Health promotion by flavonoids, tocopherols, tocotrienols, and other phenols: direct or indirect effects? Antioxidant or not? *Am. J. Clin. Nutr.* **81:** 268S–276S.
12. Fraga, C.G. 2007. Plant polyphenols: how to translate their *in vitro* antioxidant actions to *in vivo* conditions. *IUBMB Life* **59:** 308–315.

13. Holst, B. & G. Williamson. 2008. Nutrients and phytochemicals: from bioavailability to bioefficacy beyond antioxidants. *Curr. Opin. Biotechnol.* **19:** 73–82.
14. Fraga, C.G. *et al.* 2010. Basic biochemical mechanisms behind the health benefits of polyphenols. *Mol. Aspects Med.* **31:** 435–445.
15. Galleano, M. *et al.* 2010. Antioxidant actions of flavonoids: thermodynamic and kinetic analysis. *Arch. Biochem. Biophys.* **501:** 23–30.
16. Kopelman, P.G. 2000. Obesity as a medical problem. *Nature* **404:** 635–643.
17. Despres, J.P. *et al.* 2008. Abdominal obesity and the metabolic syndrome: contribution to global cardiometabolic risk. *Arterioscler. Thromb. Vasc. Biol.* **28:** 1039–1049.
18. Ross, R. *et al.* 2002. Abdominal obesity, muscle composition, and insulin resistance in premenopausal women. *J. Clin. Endocrinol. Metab.* **87:** 5044–5051.
19. Barth, S.W. *et al.* 2011. Moderate effects of apple juice consumption on obesity-related markers in obese men: impact of diet-gene interaction on body fat content. *Eur. J. Nutr.* [Epub ahead of print Oct 25]. doi: 10.1007/s00394-011-0264-6.
20. Almoosawi, S. *et al.* 2010. The effect of polyphenol-rich dark chocolate on fasting capillary whole blood glucose, total cholesterol, blood pressure and glucocorticoids in healthy overweight and obese subjects. *Br. J. Nutr.* **103:** 842–850.
21. Bell, Z.W. *et al.* 2011. A dual investigation of the effect of dietary supplementation with licorice flavonoid oil on anthropometric and biochemical markers of health and adiposity. *Lipids Health Dis.* **10:** 29.
22. Knab, A.M. *et al.* 2011. Quercetin with vitamin C and niacin does not affect body mass or composition. *Appl. Physiol. Nutr. Metab.* **36:** 331–338.
23. Phung, O.J. *et al.* 2010. Effect of green tea catechins with or without caffeine on anthropometric measures: a systematic review and meta-analysis. *Am. J. Clin. Nutr.* **91:** 73–81.
24. Dulloo, A.G. *et al.* 1999. Efficacy of a green tea extract rich in catechin polyphenols and caffeine in increasing 24-h energy expenditure and fat oxidation in humans. *Am. J. Clin. Nutr.* **70:** 1040–1045.
25. Rains, T.M. *et al.* 2011. Antiobesity effects of green tea catechins: a mechanistic review. *J. Nutr. Biochem.* **22:** 1–7.
26. Kawaguchi, K. *et al.* 1997. Hesperidin as an inhibitor of lipases from porcine pancreas and Pseudomonas. *Biosci. Biotechnol. Biochem.* **61:** 102–104.
27. Sbarra, V. *et al.* 2005. *In vitro* polyphenol effects on activity, expression and secretion of pancreatic bile salt-dependent lipase. *Biochim. Biophys. Acta* **1736:** 67–76.
28. Babu, P.V. & D. Liu. 2008. Green tea catechins and cardiovascular health: an update. *Curr. Med. Chem.* **15:** 1840–1850.
29. Peng, C.H. *et al.* 2011. Mulberry water extracts possess an anti-obesity effect and ability to inhibit hepatic lipogenesis and promote lipolysis. *J. Agric. Food Chem.* **59:** 2663–2671.
30. Goto, T. *et al.* 2011. Tiliroside, a glycosidic flavonoid, ameliorates obesity-induced metabolic disorders via activation of adiponectin signaling followed by enhancement of fatty acid oxidation in liver and skeletal muscle in obese-diabetic mice. *J. Nutr. Biochem.* [Epub ahead of print Sept 1].
31. Guzik, T.J. *et al.* 2006. Adipocytokines—novel link between inflammation and vascular function? *J. Physiol. Pharmacol.* **57:** 505–528.
32. Hotamisligil, G.S. *et al.* 1993. Adipose expression of tumor necrosis factor-alpha: direct role in obesity-linked insulin resistance. *Science* **259:** 87–91.
33. Fraga, C.G. & P.I. Oteiza. 2011. Dietary flavonoids: role of (–)-epicatechin and related procyanidins in cell signaling. *Free Radic. Biol. Med.* **51:** 813–823.
34. Vazquez-Prieto, M.A. *et al.* 2011. Effect of red wine on adipocytokine expression and vascular alterations in fructose-fed rats. *Am. J. Hypertens.* **24:** 234–240.
35. Rivera, L. *et al.* 2008. Quercetin ameliorates metabolic syndrome and improves the inflammatory status in obese Zucker rats. *Obesity (Silver Spring)* **16:** 2081–2087.
36. Bose, M. *et al.* 2008. The major green tea polyphenol, (–)-epigallocatechin-3-gallate, inhibits obesity, metabolic syndrome, and fatty liver disease in high-fat-fed mice. *J. Nutr.* **138:** 1677–1683.
37. Terra, X. *et al.* 2011. Modulatory effect of grape-seed procyanidins on local and systemic inflammation in diet-induced obesity rats. *J. Nutr. Biochem.* **22:** 380–387.
38. Sakurai, T. *et al.* 2010. Oligomerized grape seed polyphenols attenuate inflammatory changes due to antioxidative properties in coculture of adipocytes and macrophages. *J. Nutr. Biochem.* **21:** 47–54.
39. Overman, A. *et al.* 2011. Quercetin attenuates inflammation in human macrophages and adipocytes exposed to macrophage-conditioned media. *Int. J. Obes.* **35:** 1165–1172.
40. Liu, H.S. *et al.* 2006. Inhibitory effect of green tea (–)-epigallocatechin gallate on resistin gene expression in 3T3-L1 adipocytes depends on the ERK pathway. *Am. J. Physiol. Endocrinol. Metab.* **290:** E273–E281.
41. Vazquez-Prieto, M.A. *et al.* 2012. (-)-Epicatechin prevents TNFα-induced activation of signaling cascades involved in inflammation and insulin sensitivity in 3T3-L1 adipocytes. Arch. Biochem. Biophys. [Epub ahead of print March 8].
42. Imai, K. & K. Nakachi. 1995. Cross sectional study of effects of drinking green tea on cardiovascular and liver diseases. *Br. Med. J.* **310:** 693–696.
43. Basu, A. & T.J. Lyons. 2011. Strawberries,blueberries, and cranberries in the metabolic syndrome: clinical perspectives. *J. Agric. Food Chem.* [Epub ahead of print Nov 29]. doi: 10.1021/jf203488k.
44. Nagao, T. *et al.* 2009. A catechin-rich beverage improves obesity and blood glucose control in patients with type 2 diabetes. *Obesity (Silver Spring)* **17:** 310–317.
45. Egert, S. *et al.* 2010. Serum lipid and blood pressure responses to quercetin vary in overweight patients by apolipoprotein E genotype. *J. Nutr.* **140:** 278–284.
46. Mulvihill, E.E. *et al.* 2009. Naringenin prevents dyslipidemia, apolipoprotein B overproduction, and hyperinsulinemia in LDL receptor-null mice with diet-induced insulin resistance. *Diabetes* **58:** 2198–2210.
47. Yokozawa, T. *et al.* 2008. Gravinol ameliorates high-fructose-induced metabolic syndrome through regulation of lipid metabolism and proinflammatory state in rats. *J. Agric. Food Chem.* **56:** 5026–5032.

48. Shabrova, E.V. *et al.* 2011. Insights into the molecular mechanisms of the anti-atherogenic actions of flavonoids in normal and obese mice. *PLoS One* **6:** e24634.
49. Mulvihill, E.E. *et al.* 2010. Naringenin decreases progression of atherosclerosis by improving dyslipidemia in high-fat-fed low-density lipoprotein receptor-null mice. *Arterioscler. Thromb. Vasc. Biol.* **30:** 742–748.
50. Huang, H.C. & J.K. Lin. 2011. Pu-erh tea, green tea, and black tea suppresses hyperlipidemia, hyperleptinemia and fatty acid synthase through activating AMPK in rats fed a high-fructose diet. *Food Funct.* [Epub ahead of print Nov 30]. doi: 10.1039/C1FO10157A.
51. Horton, J.D. *et al.* 2002. SREBPs: activators of the complete program of cholesterol and fatty acid synthesis in the liver. *J. Clin. Invest.* **109:** 1125–1131.
52. Del Bas, J.M. *et al.* 2008. Dietary procyanidins lower triglyceride levels signaling through the nuclear receptor small heterodimer partner. *Mol. Nutr. Food Res.* **52:** 1172–1181.
53. Del Bas, J.M. *et al.* 2009. Dietary procyanidins enhance transcriptional activity of bile acid-activated FXR *in vitro* and reduce triglyceridemia *in vivo* in a FXR-dependent manner. *Mol. Nutr. Food Res.* **53:** 805–814.
54. Quesada, H. *et al.* 2009. Grape seed proanthocyanidins correct dyslipidemia associated with a high-fat diet in rats and repress genes controlling lipogenesis and VLDL assembling in liver. *Int. J. Obes.* **33:** 1007–1012.
55. Baiges, I. *et al.* 2010. Lipogenesis is decreased by grape seed proanthocyanidins according to liver proteomics of rats fed a high fat diet. *Mol. Cell Proteomics* **9:** 1499–1513.
56. Honda, K. *et al.* 2009. The molecular mechanism underlying the reduction in abdominal fat accumulation by licorice flavonoid oil in high fat diet-induced obese rats. *Anim. Sci. J.* **80:** 562–569.
57. Guo, H.X. *et al.* 2009. Long-term baicalin administration ameliorates metabolic disorders and hepatic steatosis in rats given a high-fat diet. *Acta Pharmacol. Sin.* **30:** 1505–1512.
58. Kamisoyama, H. *et al.* 2008. Investigation of the anti-obesity action of licorice flavonoid oil in diet-induced obese rats. *Biosci. Biotechnol. Biochem.* **72:** 3225–3231.
59. Lee, Y.A. *et al.* 2008. Effects of proanthocyanidin preparations on hyperlipidemia and other biomarkers in mouse model of type 2 diabetes. *J. Agric. Food Chem.* **56:** 7781–7789.
60. Galleano, M. *et al.* Hypertension, nitric oxide, oxidants, and dietary plant polyphenols. *Curr. Pharm. Biotechnol.* **11:** 837–848.
61. Cassidy, A. *et al.* 2011. Habitual intake of flavonoid subclasses and incident hypertension in adults. *Am. J. Clin. Nutr.* **93:** 338–347.
62. Davison, K. *et al.* 2008. Effect of cocoa flavanols and exercise on cardiometabolic risk factors in overweight and obese subjects. *Int. J. Obes.* **32:** 1289–1296.
63. Rizza, S. *et al.* 2011. Citrus polyphenol hesperidin stimulates production of nitric oxide in endothelial cells while improving endothelial function and reducing inflammatory markers in patients with metabolic syndrome. *J. Clin. Endocrinol. Metab.* **96:** E782–E792.
64. Basu, A. *et al.* 2011. Green tea minimally affects biomarkers of inflammation in obese subjects with metabolic syndrome. *Nutrition* **27:** 206–213.
65. Tsai, H.Y. *et al.* 2008. Effect of a proanthocyanidin-rich extract from longan flower on markers of metabolic syndrome in fructose-fed rats. *J. Agric. Food Chem.* **56:** 11018–11024.
66. Ihm, S.H. *et al.* 2009. Catechin prevents endothelial dysfunction in the prediabetic stage of OLETF rats by reducing vascular NADPH oxidase activity and expression. *Atherosclerosis* **206:** 47–53.
67. Vazquez-Prieto, M.A. *et al.* 2010. Dealcoholized red wine reverse vascular remodeling in an experimental model of metabolic syndrome: role of NAD(P)H oxidase and eNOS activity. *Food Funct.* **1:** 124–129.
68. Agouni, A. *et al.* 2009. Red wine polyphenols prevent metabolic and cardiovascular alterations associated with obesity in Zucker fatty rats (Fa/Fa). *PLoS One* **4:** e5557.
69. Banini, A.E. *et al.* 2006. Muscadine grape products intake, diet and blood constituents of non-diabetic and type 2 diabetic subjects. *Nutrition* **22:** 1137–1145.
70. Shrime, M.G. *et al.* 2011. Flavonoid-rich cocoa consumption affects multiple cardiovascular risk factors in a meta-analysis of short-term studies. *J. Nutr.* **141:** 1982–1988.
71. Song, Y. *et al.* 2005. Associations of dietary flavonoids with risk of type 2 diabetes, and markers of insulin resistance and systemic inflammation in women: a prospective study and cross-sectional analysis. *J. Am. Coll. Nutr.* **24:** 376–384.
72. Kataja-Tuomola, M.K. *et al.* Intake of antioxidants and risk of type 2 diabetes in a cohort of male smokers. *Eur. J. Clin. Nutr.* **65:** 590–597.
73. Sivakumar, A.S. & C.V. Anuradha. 2011. Effect of galangin supplementation on oxidative damage and inflammatory changes in fructose-fed rat liver. *Chem. Biol. Interact.* **193:** 141–148.
74. Cao, H. *et al.* 2007. Green tea polyphenol extract regulates the expression of genes involved in glucose uptake and insulin signaling in rats fed a high fructose diet. *J. Agric. Food Chem.* **55:** 6372–6378.
75. Qin, B. *et al.* 2010. Green tea polyphenols improve cardiac muscle mRNA and protein levels of signal pathways related to insulin and lipid metabolism and inflammation in insulin-resistant rats. *Mol. Nutr. Food Res.* **54:** S14–S23.
76. Meeprom, A. *et al.* 2011. Grape seed extract supplementation prevents high-fructose diet-induced insulin resistance in rats by improving insulin and adiponectin signalling pathways. *Br. J. Nutr.* **106:** 1173–1181.
77. Tzeng, Y.M. *et al.* 2009. Kaempferitrin activates the insulin signaling pathway and stimulates secretion of adiponectin in 3T3-L1 adipocytes. *Eur. J. Pharmacol.* **607:** 27–34.

Ann. N.Y. Acad. Sci. ISSN 0077-8923

ANNALS OF THE NEW YORK ACADEMY OF SCIENCES
Issue: *Environmental Stressors in Biology and Medicine*

Dietary polyphenols in cancer prevention: the example of the flavonoid quercetin in leukemia

Carmela Spagnuolo,[1,*] Maria Russo,[1,*] Stefania Bilotto,[1] Idolo Tedesco,[1] Bruna Laratta,[2] and Gian Luigi Russo[1]

[1]Istituto di Scienze dell'Alimentazione, Consiglio Nazionale delle Ricerche, Avellino, Italy. [2]Istituto di Chimica Biomolecolare, Consiglio Nazionale delle Ricerche, Pozzuoli (Napoli), Italy

Address for correspondence: Gian Luigi Russo, Istituto Scienze dell'Alimentazione, Via Roma, 64, 83100 Avellino, Italy. glrusso@isa.cnr.it

Increased consumption of fruit and vegetables can represent an easy strategy to significantly reduce the incidence of cancer. We recently demonstrated that the flavonoid quercetin, naturally present in the diet and belonging to the class of phytochemicals, is able to sensitize several leukemia cell lines and B cells isolated from patients affected by chronic lymphocytic leukemia (B-CLL), in addition to apoptotic inducers (anti-CD95 and rTRAIL). Further, it potentiates the effect of fludarabine, a first-line chemotherapeutic drug used against CLL. The proapoptotic activity of quercetin in cell lines and B-CLL is related to the expression and activity of Mcl-1–antiapoptotic proteins belonging to the Bcl-2 family. Quercetin downregulates Mcl-1 mRNA and protein levels acting on mRNA stability and protein degradation. Considering the low toxicity of the flavonoids toward normal peripheral blood cells, our experimental results are in favor of a potential use of quercetin in adjuvant chemotherapy in CLL or other types of cancer.

Keywords: quercetin; chemoprevention; chronic lymphocytic leukemia; phytochemicals; apoptosis

Introduction

A strong functional correlation exists between diet and cancer. An incorrect diet may influence the occurrence of overall types of cancer from 10% to 70%.[1] A large number of epidemiological studies suggest that a daily intake of fruits and vegetables can reduce the incidence of several types of cancers.[2–9]

Fruits and vegetables contain small molecules, named phytochemicals, showing *in vitro*–and *in vivo*–specific anticancer properties. Phytochemicals are organic components of plants not "essential" for life.[2,10] They possess a variety of physiological functions in plant tissues, such as regulation of metabolic enzymes and defense against xenobiotics. Phytochemicals attracted the interest of scientists because of the demonstration that their biological targets in mammalian cells are the same as those involved in inflammatory processes and oncogenic transformation—key factors such as controlling the cell cycle, apoptosis evasion, angiogenesis, and metastases. The study of the anticancer activities of phytochemicals has contributed to the field of chemoprevention, which is defined as the use of naturally occurring agents to inhibit, reverse, or retard tumorigenesis.[10] Chemopreventive agents act on signal transduction regulation at different levels by modulating hormone/growth factor activity, inhibiting oncogene activity and activating tumor suppressor genes, inducing terminal differentiation, activating apoptosis, restoring immune response, inhibiting angiogenesis, decreasing inflammation, and scavenging reactive oxygen species (ROS).[11] They can be divided into two main categories: blocking and suppressing agents. The former prevent carcinogens from damaging cells by several mechanisms, including enhancing detoxification, modifying carcinogen uptake and metabolism,

*These two authors contributed equally to the preparation of the present work.

doi: 10.1111/j.1749-6632.2012.06599.x

scavenging ROS and other oxidative species, and enhancing DNA repair.[11–13] The latter inhibit cancer promotion and progression after the formation of preneoplastic cells.[13]

In recent years, we have focused our attention on the activity of quercetin, one of the main flavonoids present in fruits and vegetables.[14] This class of polyphenols contains a basic skeleton of diphenylpropane C_6–C_3–C_6 that is widely present in plant-derived foods and beverages. The uptake of quercetin ranges from between 5 mg/day and 40 mg/day, with an increase up to 10-fold if the diet includes fruits and vegetables particularly rich in this compound, such as onions, apples, and strawberries. As we have recently reviewed,[14] at chemopreventive or pharmacological doses, quercetin has been shown to modulate almost all of the different hallmarks of cancer as defined by Hanahan and Weinberg.[15,16] However, it has also been suggested that as an alternative to its potential as a drug, quercetin, like other polyphenols, can be used as an active component in the formulation of dietary supplements or nutraceuticals.[17] In fact, several examples have been reported in the scientific literature demonstrating a synergistic interaction between quercetin and other bioactive polyphenols. For example, in the MOLT-4 human leukemia cell line, the interaction of ellagic acid and quercetin synergistically enhanced the anticarcinogenic potential of the individual compounds.[18] Ellagic acid potentiated the effects of quercetin for $p21^{cip1/waf1}$ protein levels and p53 phosphorylation at serine 15, possibly explaining the synergistic effect observed in apoptosis induction.[19] The extensive methylation of green tea polyphenols *in vivo* may limit their chemopreventive potential. The combination of quercetin and epigallocatechin gallate decreases the methylation rate of epigallocatechin gallate and increased its cellular absorption.[20] In the human pancreatic carcinoma cell line Mia PACA-2, resveratrol more than doubled the cytochrome c release and caspase-3 activity caused by quercetin alone, suggesting that these two agents act on the mitochondrial permeability transition pore through distinct pathways.[21] However, in human leukemia cells, the synergistic effects of quercetin and resveratrol in the induction of apoptosis and transient cell cycle arrest was observed only after a delay (>48 hours).[22] Finally, pterostilbene, a naturally occurring analog of resveratrol, administrated intravenously in combination with quercetin to mice inhibited 73% of the metastatic growth of B16M-F10 cells in the liver, a common site for metastasis development.[23] The synergistic effects of quercetin, resveratrol, and pterostilbene have also been shown to be effective in the protection of erythrocyte membranes against lipid peroxidation. At lower concentrations, resveratrol with quercetin or pterostilbene synergistically inhibited the oxidative injury of membrane lipids. At higher concentrations, an additive effect was observed.[24]

Here, we will describe that differentiation among uses of quercetin depends upon applied doses, bioavailability, and pleiotropic effects.

Quercetin metabolism, bioavailability, and toxicity

The chemopreventive effect of polyphenols has often been associated with their antioxidant activity. Quercetin acts as a powerful antioxidant, since H_2O_2-induced ROS increase in cell lines is lowered by the addition of micromolar concentration of quercetin.[25] However, the antioxidant properties of quercetin are not directly correlated with its ability to inhibit cell growth.[25–27]

After absorption, quercetin is metabolized in different organs, such as the small intestine, colon, liver, and kidney. Here, the molecule is conjugated to methyl and sulfate groups and glucuronic acid.[28] In addition, the absorption of quercetin is influenced by gut microflora. As a result of its absorption and metabolism, total quercetin derived from the diet is present in plasma at the nanomolar range (<100 nM), but can be increased to micromolar concentrations after supplementation. For example, 28 days of supplementation with 1 g/day of quercetin increased plasma concentrations to 1.5 μM.[29,30] However, the critical review of a significant number of works regarding the antiproliferative activity of flavonoids in cell lines of different origins evidences that the minimal concentration required to reduce cell viability is above 5 μM, a value significantly far from the nutritional polyphenol concentrations.[30,31] This suggests that it is very difficult to predict the healthy effects of quercetin and other polyphenols only based on *in vitro* studies. In addition, phytochemical bioavailability also influences another important phenomenon largely demonstrated on cell lines, for example, the hormetic effect.[32] Hormesis is defined as the ability of a substance, when administered at low doses, to exert an opposite action than that seen at high

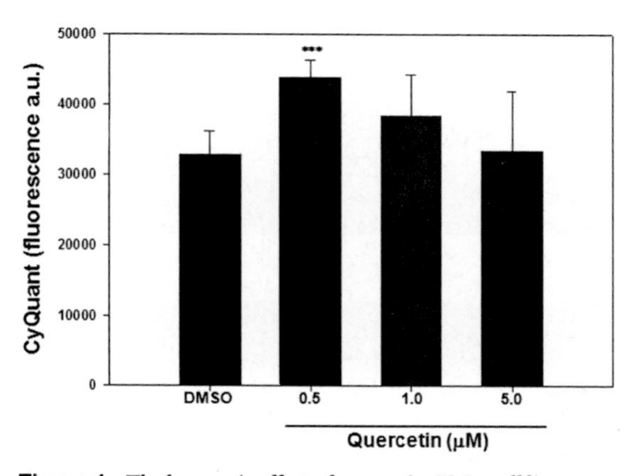

Figure 1. The hormetic effect of quercetin. HeLa cell lines were treated with 0.5–5 μM of quercetin for 48 h and cell proliferation was assessed by a CyQUANT Direct Cell Proliferation Assay Kit. Bar graphs represent the mean ± SD. Symbols indicate significance with $P < 0.05$ (***) respect to DMSO treated cells.

concentrations. As reported in Figure 1, a significant increase in cell growth was observed when quercetin was applied at low concentrations (0.5 μM), comparable to those reachable *in vivo* after a meal abundant in quercetin-enriched foods. For these reasons, it is necessary to carefully consider the use of phytochemicals as supplements or functional foods, since potential benefits may become damaging if precancer cells are affected by the hormetic effect.

Although the International Agency for Research on Cancer (IARC) has stated that quercetin cannot be classifiable as carcinogenic to humans,[33] a study carried out by the National Toxicology Program (NTP) on male F344/N rats fed with 2 g/kg body weight/day of quercetin showed severe chronic nephropathy, hyperplasia, and neoplasia of the renal tubular epithelium.[28] More recently, it has been reported that maternal intake of quercetin during gestation could increase the risk on mixed-lineage leukemia (MLL) gene rearrangements, which are frequently observed in childhood leukemia, especially in the presence of compromised DNA repair.[34] The same group reported that offspring that had received quercetin during gestation had lowered benzo[α]pyrene-induced DNA adduct formation,[35] and had a significant increase in iron storage and an increased expression of inflammation-associated cytokines in the liver.[36] In the first case, the authors suggested that prenatal exposure to quercetin could decrease the risk for benzo[α]pyrene-induced DNA damage and therefore the susceptibility to lung cancer.[35] Accordingly, others have demonstrated that orally administered quercetin (up to 2 g/kg body weight) did not induce unscheduled DNA synthesis in hepatocytes of male or female rats, suggesting that quercetin was not genotoxic.[37]

In humans, the unique phase I clinical trial of quercetin recommended a dose of 1.4 g/m^2, which corresponds to about 2.5 g for a 70 kg individual, administered via intravenous infusion at three-week or weekly intervals. At higher doses, up to 50 mg/kg (about 3.5 g/70 kg), renal toxicity was detected without signs of nephritis or obstructive uropathy.[38]

Quercetin and apoptosis

The only way to bypass the problem of low bioavailability of quercetin and other polyphenols is the administration of these compounds at pharmacological doses, in order to reach the concentrations needed to kill cancer cells. We now concentrate our attention on the ability of quercetin to sensitize malignant cell lines to programmed cell death when associated with apoptotic inducers. Apoptosis is targeted by many phytochemicals and is a fundamental process to eliminate premalignant and malignant cells. In physiological conditions, apoptosis can follow two different pathways: the intrinsic and the extrinsic pathways.[39,40] The former is controlled at the mitochondrial level and is activated by genotoxic damage, which prevents cells from replicating abnormally. In this case, Bcl-2 family members, including factors possessing proapoptotic (Bax, Bak, PUMA, NOXA, Bim, Bid) or antiapoptotic activities (Bcl-2, Mcl-1, Bcl-xL, Bfl-1/A1)[40] are key players. The extrinsic pathway is induced by extracellular ligands, such as CD95 and TRAIL, which bind to specific receptors on the membrane, for example, death receptors (DRs). This interaction causes receptor trimerization and formation of a complex called DISC (death-inducing signaling complex), which triggers a cascade of apoptotic signals culminating in the activation of a class of cysteine proteases (caspases) that lead to the destruction and elimination of tumor cells without inflammation and tissue damage.[39] One of the goals of traditional chemotherapy is to induce apoptosis in cancer cells. Promising results are expected from ongoing phases I/II clinical trials where treatments are based on DR agonistic antibodies and/or recombinant proteins administered alone or in association with classic and novel chemotherapeutic drugs.[39] However, many tumors are, or become, resistant to

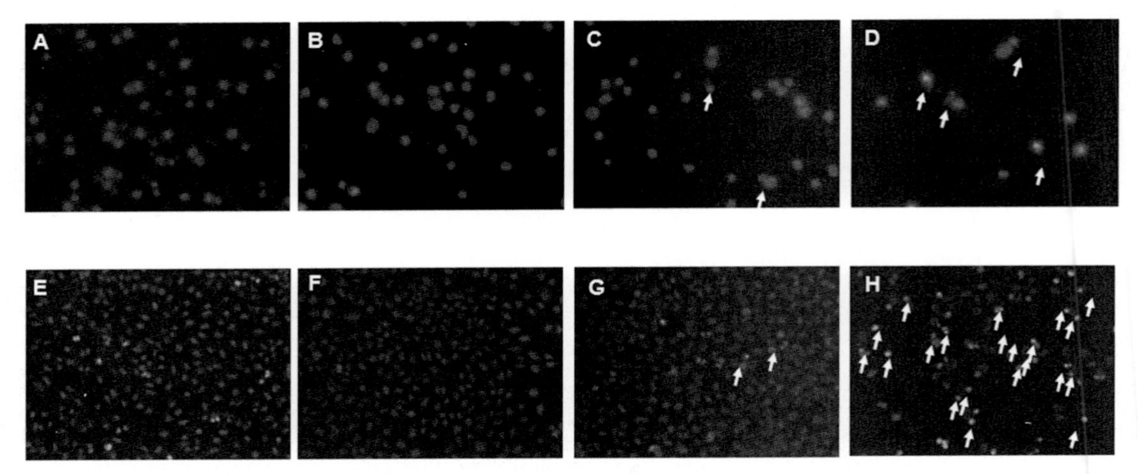

Figure 2. The enhancing apoptotic effect of quercetin in HPB-ALL (panels A–D) and U2Os (panels E–H) cell lines. Cells were treated for 24 h with 50 μM quercetin (panels B–F), 50 ng/mL anti-CD95 (panel C), or 10 ng/mL rTRAIL (recombinant TRAIL; panel G). The association between quercetin and anti-CD95 (panel D) or rTRAIL (panel H). DMSO 0.1% was used as a control (panels A–E). Chromatin condensation and formation of apoptotic bodies was revealed by DAPI nuclear staining. White arrows indicate apoptotic cells.

therapeutic protocols based on apoptotic inducers. We have demonstrated the ability of quercetin to enhance apoptosis induced by anti-CD95 and recombinant TRAIL (rTRAIL) in HPB-ALL, an apoptosis-resistant leukemic cell line (Fig. 2).[25,26,41] It is important to emphasize that quercetin per se was neither apoptotic nor cytotoxic, and its effect was specific and independent of the antioxidant properties of the molecule; in fact, other dietary flavonols structurally and functionally related to quercetin (e.g., catechin and myricetin) did not mimic the apoptogenic activity of quercetin. However, these analogs did maintain their antioxidant capacity.[26] This finding has been confirmed in other cellular models for human leukemia of lymphoid or myeloid origin.[41] It is worthwhile mentioning that quercetin in the concentration range applied to cancer cell lines did not induce apoptosis in the same cell lines induced to differentiate or in lymphocytes isolated from the peripheral blood of healthy subjects.[27] This selectivity of quercetin encourages a potential use of the molecule as a coadjuvant in those chemotherapeutic protocols where the lack of specificity toward noncancer cells is an unwanted and deleterious collateral effect.

Quercetin in chronic lymphocytic leukemia

The growth-suppressive effects of quercetin in leukemic cells was observed early in the 1990s by the pivotal works of Larocca *et al.* who demonstrated that, in a dose-dependent manner, the molecule was able to inhibit the *in vitro* growth of isolated blasts from acute myeloid leukemias (AML) and acute lymphoid leukemias (ALL) and to enhance the antiproliferative effects of cytosine arabinoside and hyperthermia.[42–44] More recently, it has been reported that quercetin suppressed the activity of telomerase in ALL and chronic myeloid leukemia (CML) cells,[45] induced Fas ligand–related apoptosis in human leukemia HL-60 cells by promotion of histone H3 acetylation,[46] decreased the level of Notch1 protein and its active fragment in the lymphoblastic leukemia cell line with constitutive Notch activation by acting as a Wnt inhibitor,[47] inhibited murine leukemia WEHI-3 cells when injected into BALB/c mice, and promoted macrophage phagocytosis and natural killer cell activity.[48]

We have recently focused our interest on the sensitizing effect of quercetin toward DR-induced cell death in B cells isolated from chronic lymphocytic patients (B-CLL). This type of leukemia has been selected because B-CLL is one of the most common leukemias in the adult population (22–30% of all leukemia cases);[49] it is a malignancy of mature B lymphocytes with a low proliferative rate and apoptosis resistance, mainly to death ligand (DL) stimuli; and it is a highly heterogeneous and still incurable disease—most treated patients

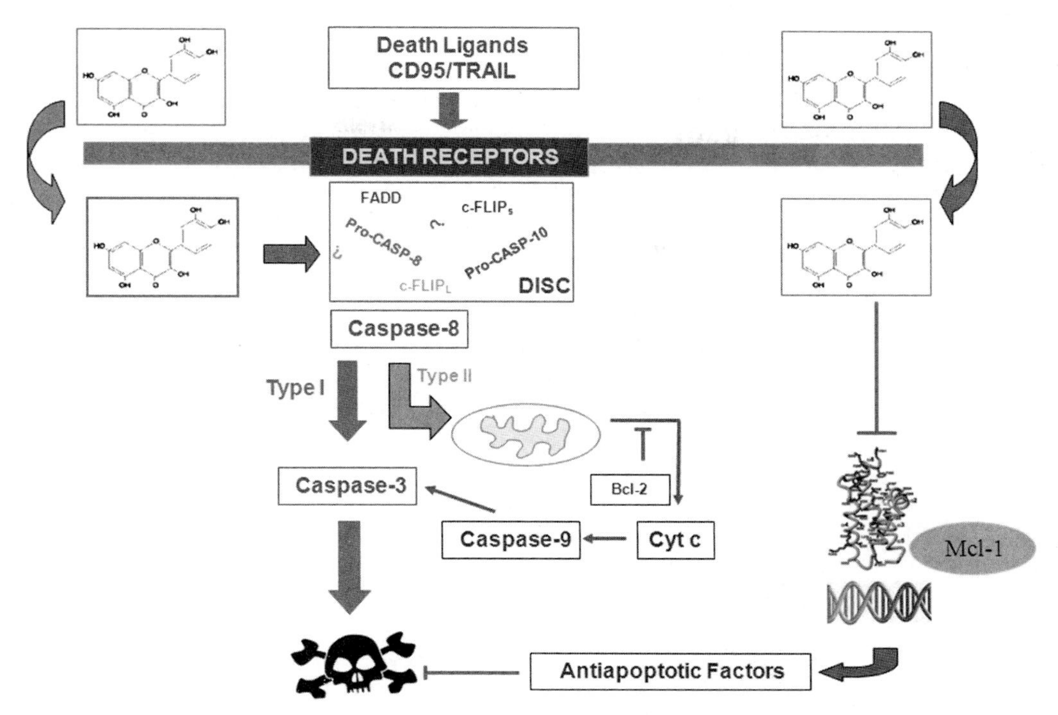

Figure 3. Pleiotropic effects of quercetin in enhancing apoptosis. Quercetin can reduce or bypass resistance to DL-induced apoptosis, acting on inhibitor(s) located at the DISC level and acting on both type-I and type-II cells.[61] In addition, quercetin is able to downregulate antiapoptotic factor Mcl-1, interfering with its mRNA stability and protein degradation (see the text for description).

became resistant to common chemotherapeutic drugs (fludarabine).

We isolated mononuclear cells from the peripheral blood of CLL patients, selecting those patients whose sensitivity to quercetin was low and resistance to apoptosis induced by anti-CD95 and rTRAIL was significantly high according to data reported in the literature.[50,51] In these selected samples, we observed that quercetin enhanced sensitivity to anti-CD95 and rTRAIL treatment with an increase in cell death of about 1.5- and 1.6-fold, respectively, when compared with quercetin monotreatment.[27] However, due to the well-known heterogeneity of B-CLL, 10% of samples did not show any response to quercetin alone or in combination with DL agonists (anti-CD95 and rTRAIL).

Based on these observations, we speculate that quercetin may have the capacity to strengthen the efficacy of drugs widely used in the therapy of CLL, such as fludarabine.[52] In fact, B cells isolated from CLL patients and resistant to 3–14 μM fludarabine (a range of values corresponding to the therapeutic plasma concentration of fludarabine[53]) showed an increase in cell death of approximately twofold in combined treatment when compared with quercetin and by sixfold with respect to fludarabine.[27] We have also demonstrated that in B cells isolated from two different patients (CLL56 and CLL42 in Ref. 27), showing a different grade of resistance to DL agonists and fludarabine, that resistance to anti-CD95 and rTRAIL was bypassed by quercetin in CLL56, but not in CLL42. Here, significant cell death was achieved by combined treatment with fludarabine and quercetin.[27] Overall, our data demonstrate that the presence of quercetin ameliorates sensitivity of B-CLL to apoptosis in the presence of both traditional therapeutic drugs (e.g., fludarabine) and new protocols based on DR agonists.

This conclusion represents a good example of the suggested pleiotropic effect of quercetin: when the molecule is not able to bypass resistance to DLs in B-CLL samples, it increases the cell death-ameliorating cytotoxic effect of fludarabine, and vice versa.

The mechanism by which quercetin acts in leukemic cell lines and B-CLL cells is still a matter of study. Quercetin, as a pleiotropic molecule, may exert its effects on different pathways (Fig. 3). We have demonstrated in leukemic cell lines[25,41] and in B cells from CLL patients[27] that the well-known antioxidant properties of the molecule are independent of its apoptogenic activity when associated with DLs or fludarabine. In fact, changes of intracellular ROS do not correlate with quercetin treatments and/or its cytotoxicity. Moreover, to explain the ability of the molecule to sensitize the cancer cells to rTRAIL-induced apoptosis, we hypothesized that quercetin may be able to stimulate the expression of TRAIL receptor(s). However, the expression of TRAIL-R1 (predominantly expressed in B-CLL over TRAIL-R2) fluctuates among samples, and quercetin treatment may induce significant increase in TRAIL-R1 expression. Therefore, we exclude the possibility that the sensitizing effect of quercetin in rTRAIL-induced apoptosis could be associated to an increased number of receptor molecules on the cell membrane.[27]

Alternatively, considering the importance of Bcl-2 family members in controlling apoptosis in B-CLL,[54] we tested whether quercetin can interfere with Bcl-2 protein regulation. Our interest was essentially focused on Bcl-2 and Mcl-1 proteins, which are particularly important in apoptosis regulation in B-CLL and also correlated with its resistance to common therapies.[27] In the case of prosurvival member Bcl-2, which is highly expressed in B-CLL isolated from patients, its expression was not influenced by quercetin, and its association with DL agonists (anti-CD95 and rTRAL) did not decrease the Bcl-2 level. Therefore, we investigated whether the sensitizing effect of quercetin in inducing apoptosis could be associated with the modulation of expression or stability of Mcl-1. Often the high level of expression of this protein is correlated with chemotherapeutic resistance.[55] Interestingly, in B cells isolated from CLL patients expressing significant levels of Mcl-1, quercetin downregulated Mcl-1 expression in a range of concentrations (10–20 μM), which was not cytotoxic for B cells, suggesting that the molecule may interfere with Mcl-1 regulation.[56] These data were corroborated by experiments performed on U-937, a human myelomonocytic cell line, expressing a high level of Mcl-1. Here, quercetin strongly reduced Mcl-1 protein expression without affecting cell viability after one to four hours of treatment. Since this effect paralleled with reduced Mcl-1 mRNA levels, we suggest that regulation may occur at the transcriptional level or affect mRNA stability. In support of this hypothesis, inhibition of transcription was identified—inhibitition of transcription by actinomycin D in the presence of quercetin caused a significant acceleration in Mcl-1 mRNA degradation, with a 50% reduction in 135 minutes when only actinomycin D was applied compared to a 50% decrease in 79 min when actinomycin D was associated with quercetin. These results indicate the ability of the molecule to interfere with one or more processes regulating translation.[56]

To further sustain the pleiotropic effect of quercetin, we demonstrated that quercetin also interferes with posttranscriptional regulation of Mcl-1, which degrades under the control of proteasome- and/or caspase-dependent mechanisms.[57,58] In U-937, cotreatment with quercetin and Z-Vad-FMK, a cell-permeable caspase inhibitor, resulted in increased Mcl-1 protein levels, indicating its independence from caspase-mediated degradation. On the contrary, quercetin plus MG132, a proteasome inhibitor, did not revert the effect of MG132 monotreatment, strongly suggesting that quercetin positively regulates the proteasome-dependent degradation of Mcl-1.[56]

Conclusions

Quercetin appears to be a molecule possessing multiple properties all directed to reduce cell growth in cancer cells.[14] In summary, quercetin sensitizes synergistically induced cell death in human malignant cell lines to CD95/TRAIL and enhances apoptosis in B cells isolated from CLL patients resistant to DR- and fludarabine-mediated cell death. However, key issues, such as uptake, metabolism, and circulating concentrations of quercetin and its metabolites suggest that a regular diet cannot provide adequate amounts of quercetin (<1 μM) compatible with any described chemopreventive effects, although it is relatively easy to increase total quercetin concentrations in plasma by supplementation with quercetin-enriched foods or supplements. Data from animal studies suggest that concentrations of quercetin above 10 μM are attainable with oral doses.[59] In this case, the efficacy of the molecule cannot be measured in healthy individuals, but only on subjects affected by diseases.[14] The potential clinical applications change, however, when shifting from nutraceutical/supplement doses (a few milligrams diluted in the diet) to pharmacological doses (hundreds of milligrams in concentrated doses). Low concentrations do not saturate metabolic pathways; therefore, circulating, unconjugated molecules are not found in the blood. This fate is common to the majority of dietary polyphenols.[31,60] Alternatively, quercetin can be administered by intravenous injection to avoid the formation of conjugates that can reduce the bioavailability of the active moiety and dramatically alter their pharmacological properties. In this regard, the possibility that quercetin may improve efficacy of chemotherapy is strengthened by the safety of the molecule administered through intravenous infusion.[38] Human studies have failed to show any adverse effects associated with the oral administration of quercetin in a single dose of up to 4 g or after one month of 500 mg twice daily.[59] This observation represents an important issue in favor of the potential use of quercetin in combination with other therapeutic treatments that are already approved and used in clinics.

We have discussed here and elsewhere[14] that quercetin is a molecule functionally pleiotropic, possessing multiple intracellular targets, affecting different cell signaling processes usually altered in cancer cells, with limited toxicity on normal cells (Fig. 3). Simultaneously targeting multiple pathways may help to kill cancer cells and slow drug resistance onset. In addition, if the association of quercetin with improved chemotherapy or radiotherapy is demonstrated, clinicians may take advantage of the synergic effects of the combined protocols, resulting in the possibility to lower doses and, consequently, reduced toxicity. New clinical trials are absolutely necessary to ascertain the full chemopreventive and chemotherapeutic efficacy of quercetin in adjuvant cancer therapy.

Acknowledgments

C.S. was partially supported by a grant from the "Innovative approaches in the evaluation and prevention of food exposure to contaminating, persistent, and emergent toxins through the study of diet and the debugging of innovative methods for surveying project," granted by the Italian Ministry of Health/Istituto Zooprofilattico Sperimentale del Mezzogiorno.

Conflicts of interest

The authors declare no conflicts of interest.

References

1. Riboli, E. & T. Norat. 2003. Epidemiologic evidence of the protective effect of fruit and vegetables on cancer risk. *Am. J. Clin. Nutr.* **78:** 559S–569S.
2. Surh, Y.J. 2003. Cancer chemoprevention with dietary phytochemicals. *Nat. Rev. Cancer* **3:** 768–780.
3. D'Incalci, M., W.P. Steward & A.J. Gescher. 2005. Use of cancer chemopreventive phytochemicals as antineoplastic agents. *Lancet Oncol.* **6:** 899–904.
4. Aune, D. *et al.* 2011. Nonlinear reduction in risk for colorectal cancer by fruit and vegetable intake based on meta-analysis of prospective studies. *Gastroenterology* **141:** 106–118.
5. Boffetta, P. *et al.* 2010. Fruit and vegetable intake and overall cancer risk in the European Prospective Investigation into Cancer and Nutrition (EPIC). *J. Natl. Cancer Inst.* **102:** 529–537.
6. Crowe, F.L. *et al.* 2011. Fruit and vegetable intake and mortality from ischaemic heart disease: results from the European Prospective Investigation into Cancer and Nutrition (EPIC)—Heart study. *Eur. Heart J.* **32:** 1235–1243.
7. Dauchet, L. *et al.* 2006. Fruit and vegetable consumption and risk of coronary heart disease: a meta-analysis of cohort studies. *J. Nutr.* **136:** 2588–2593.
8. McCullough, M.L. *et al.* 2011. Following cancer prevention guidelines reduces risk of cancer, cardiovascular disease, and all-cause mortality. *Cancer Epidemiol. Biomarkers Prev.* **20:** 1089–1097.
9. Van Duyn, M.A. & E. Pivonka. 2000. Overview of the health benefits of fruit and vegetable consumption for the dietetics professional: selected literature. *J. Am. Diet Assoc.* **100:** 1511–1521.
10. Sporn, M.B. & N. Suh. 2002. Chemoprevention: an essential approach to controlling cancer. *Nat. Rev. Cancer* **2:** 537–543.
11. Kelloff, G.J. *et al.* 2006. Progress in chemoprevention drug development: the promise of molecular biomarkers for prevention of intraepithelial neoplasia and cancer—a plan to move forward. *Clin. Cancer Res.* **12:** 3661–3697.
12. Russo, M. *et al.* 2005. Dietary phytochemicals in chemoprevention of cancer. *Curr. Med. Chem. Immunol. Endocr. Metabol. Agents* **5:** 61–72.
13. Johnson, I.T. 2007. Phytochemicals and cancer. *Proc. Nutr. Soc.* **66:** 207–215.
14. Russo, M. *et al.* 2012. The flavonoid quercetin in disease prevention and therapy: facts and fancies. *Biochem. Pharmacol.* **83:** 6–15.
15. Hanahan, D. & R.A. Weinberg. 2000. The hallmarks of cancer. *Cell* **100:** 57–70.
16. Hanahan, D. & R.A. Weinberg. 2011. Hallmarks of cancer: the next generation. *Cell* **144:** 646–674.
17. Boots, A.W., G.R. Haenen & A. Bast. 2008. Health effects of quercetin: from antioxidant to nutraceutical. *Eur. J. Pharmacol.* **585:** 325–337.
18. Mertens-Talcott, S.U., S.T. Talcott & S.S. Percival. 2003. Low concentrations of quercetin and ellagic acid synergistically influence proliferation, cytotoxicity and apoptosis in MOLT-4 human leukemia cells. *J. Nutr.* **133:** 2669–2674.
19. Mertens-Talcott, S.U. *et al.* 2005. Ellagic acid potentiates the effect of quercetin on p21waf1/cip1, p53, and MAP-kinases

without affecting intracellular generation of reactive oxygen species *in vitro*. *J. Nutr.* **135:** 609–614.
20. Wang, P., D. Heber & S.M. Henning. 2012. Quercetin increased bioavailability and decreased methylation of green tea polyphenols *in vitro* and *in vivo*. *Food Funct*. Mar 22. [Epub ahead of print].
21. Mouria, M. *et al.* 2002. Food-derived polyphenols inhibit pancreatic cancer growth through mitochondrial cytochrome C release and apoptosis. *Int. J. Cancer* **98:** 761–769.
22. Mertens-Talcott, S.U. & S.S. Percival. 2005. Ellagic acid and quercetin interact synergistically with resveratrol in the induction of apoptosis and cause transient cell cycle arrest in human leukemia cells. *Cancer Lett.* **218:** 141–151.
23. Ferrer, P. *et al.* 2005. Association between pterostilbene and quercetin inhibits metastatic activity of B16 melanoma. *Neoplasia* **7:** 37–47.
24. Mikstacka, R., A.M. Rimando & E. Ignatowicz. 2010. Antioxidant effect of trans-resveratrol, pterostilbene, quercetin and their combinations in human erythrocytes *in vitro*. *Plant Foods Hum. Nutr.* **65:** 57–63.
25. Russo, M. *et al.* 1999. Quercetin and anti-CD95(Fas/Apo1) enhance apoptosis in HPB-ALL cell line. *FEBS Lett.* **462:** 322–328.
26. Russo, M. *et al.* 2003. Flavonoid quercetin sensitizes a CD95-resistant cell line to apoptosis by activating protein kinase Calpha. *Oncogene* **22:** 3330–3342.
27. Russo, M. *et al.* 2010. Quercetin induced apoptosis in association with death receptors and fludarabine in cells isolated from chronic lymphocytic leukaemia patients. *Br. J. Cancer* **103:** 642–648.
28. Harwood, M. *et al.* 2007. A critical review of the data related to the safety of quercetin and lack of evidence of *in vivo* toxicity, including lack of genotoxic/carcinogenic properties. *Food Chem. Toxicol.* **45:** 2179–2205.
29. Conquer, J.A. *et al.* 1998. Supplementation with quercetin markedly increases plasma quercetin concentration without effect on selected risk factors for heart disease in healthy subjects. *J. Nutr.* **128:** 593–597.
30. Manach, C. *et al.* 2005. Bioavailability and bioefficacy of polyphenols in humans: I. Review of 97 bioavailability studies. *Am. J. Clin. Nutr.* **81:** 230S–242S.
31. Scalbert, A. *et al.* 2011. Databases on food phytochemicals and their health-promoting effects. *J. Agric. Food Chem.* **59:** 4331–4348.
32. Tedesco, I., M. Russo & G.L. Russo. 2010. Commentary on 'resveratrol commonly displays hormesis: occurrence and biomedical significance'. *Hum. Exp. Toxicol.* **29:** 1029–1031.
33. IARC. 1999. Quercetin. IARC monographs on the evaluation of carcinogenic risks to human, Vol. 73: 497–515.
34. Vanhees, K. *et al.* 2011. Prenatal exposure to flavonoids: implication for cancer risk. *Toxicol. Sci.* **120:** 59–67.
35. Vanhees, K. *et al.* 2012. Maternal intake of quercetin during gestation alters ex vivo benzo[a]pyrene metabolism and DNA adduct formation in adult offspring. *Mutagenesis*. First published online: February 14, 2012.
36. Vanhees, K. *et al.* 2011. Maternal quercetin intake during pregnancy results in an adapted iron homeostasis at adulthood. *Toxicology* **290:** 350–358.
37. Utesch, D. *et al.* 2008. Evaluation of the potential *in vivo* genotoxicity of quercetin. *Mutat. Res.* **654:** 38–44.
38. Ferry, D.R. *et al.* 1996. Phase I clinical trial of the flavonoid quercetin: pharmacokinetics and evidence for *in vivo* tyrosine kinase inhibition. *Clin. Cancer Res.* **2:** 659–668.
39. Russo, M. *et al.* 2010. Exploring death receptor pathways as selective targets in cancer therapy. *Biochem. Pharmacol.* **80:** 674–682.
40. Taylor, R.C., S.P. Cullen & S.J. Martin. 2008. Apoptosis: controlled demolition at the cellular level. *Nat. Rev. Mol. Cell Biol.* **9:** 231–241.
41. Russo, G.L. 2007. Ins and outs of dietary phytochemicals in cancer chemoprevention. *Biochem. Pharmacol.* **74:** 533–544.
42. Larocca, L.M. *et al.* 1997. Differential sensitivity of leukemic and normal hematopoietic progenitors to the killing effect of hyperthermia and quercetin used in combination: role of heat-shock protein-70. *Int. J. Cancer* **73:** 75–83.
43. Larocca, L.M. *et al.* 1996. Quercetin and the growth of leukemic progenitors. *Leuk. Lymphoma.* **23:** 49–53.
44. Larocca, L.M. *et al.* 1995. Quercetin inhibits the growth of leukemic progenitors and induces the expression of transforming growth factor-beta 1 in these cells. *Blood* **85:** 3654–3661.
45. Avci, C.B. *et al.* 2011. Quercetin-induced apoptosis involves increased hTERT enzyme activity of leukemic cells. *Hematology* **16:** 303–307.
46. Lee, W.J., Y.R. Chen & T.H. Tseng. 2011. Quercetin induces FasL-related apoptosis, in part, through promotion of histone H3 acetylation in human leukemia HL-60 cells. *Oncol. Rep.* **25:** 583–591.
47. Kawahara, T. *et al.* 2009. Cyclopamine and quercetin suppress the growth of leukemia and lymphoma cells. *Anticancer Res.* **29:** 4629–4632.
48. Yu, C.S. *et al.* 2010. Quercetin inhibited murine leukemia WEHI-3 cells *in vivo* and promoted immune response. *Phytother. Res.* **24:** 163–168.
49. Yee, K.W. & S.M. O'Brien. 2006. Chronic lymphocytic leukemia: diagnosis and treatment. *Mayo Clin. Proc.* **81:** 1105–1129.
50. MacFarlane, M. *et al.* 2002. Mechanisms of resistance to TRAIL-induced apoptosis in primary B cell chronic lymphocytic leukaemia. *Oncogene* **21:** 6809–6818.
51. MacFarlane, M. *et al.* 2005. Chronic lymphocytic leukemic cells exhibit apoptotic signaling via TRAIL-R1. *Cell Death Differ.* **12:** 773–782.
52. Schnaiter, A. & S. Stilgenbauer. 2010. Refractory chronic lymphocytic leukemia—new therapeutic strategies. *Oncotarget* **1:** 472–482.
53. Binet, J.L. 1993. Fludarabine phosphate in chronic lymphoproliferative diseases. The French Group on CLL. *Nouv. Rev. Fr. Hematol.* **35:** 5–7.
54. Packham, G. & F.K. Stevenson. 2005. Bodyguards and assassins: Bcl-2 family proteins and apoptosis control in chronic lymphocytic leukaemia. *Immunology* **114:** 441–449.
55. Pepper, C. *et al.* 2008. Mcl-1 expression has *in vitro* and *in vivo* significance in chronic lymphocytic leukemia and is associated with other poor prognostic markers. *Blood* **112:** 3807–3817.

rapidly, with areas of fibrosis with TGF-β expression, whereas C57BL/6J mice show extensive goblet cell hyperplasia, with expression of MUC5AC and IL-4 and IL-13 in the airways.[66]

Mouse models have also been largely used to study the role of oxidative stress in CS-induced damage. An imbalance between oxidants and antioxidants plays an important role in acute and chronic CS-induced airway diseases.[36,42,72] *Ex vivo* experiments have demonstrated that C57BL/6J alveolar macrophages show basal levels of intracellular ROS and H_2O_2 significantly higher than resistant strain ICR.[58] Conversely, C57BL/6J alveolar macrophages showed lower baseline levels of GSH compared with ICR. Such an imbalance has been shown to occur in C57BL/6J mice also *in vivo*, where the decreased level of GSH was associated with the increased level of lipid peroxidation in susceptible mouse strains.[56]

An additional tool for studying the effects of pollutants on lung is provided by the use of knockout mice. TGF-β is known to play a complex role in COPD. Mice deficient in integrin avb6 fail to activate latent TGF-β in the lungs, and mice develop inflammation with excessive metalloprotease (MMP)-12 leading to spontaneous development of emphysema.[73] ETS and mainstream CS exposure in mice is associated with MMP-12 expression in macrophages and this protease has been shown to underlie cigarette-induced emphysema in mice.[70,74] Overexpression of IL-18 increased interferon-γ (IFN-γ and IL-13 and led to severe emphysematous changes, with and chronic inflammatory changes characteristic of COPD.[75] Similarly, overexpression of IL-13 and IFN-γ in mice resulted in inflammation and lung destruction that was MMP-9 and MMP-12–dependent,[76,77] and goblet cell hyperplasia and subepithelial collagen deposition was observed. Conversely, mainstream CS-induced inflammation and emphysema were inhibited using a knockout model of IL-18 receptor.[78] The role of TNF-α in COPD was tested in TNF-α receptor knockout mice. After chronic ETS exposure, there was partial inhibition of MMP-12 elevation, neutrophil and macrophage recruitment, and alveolar enlargement.[79]

Knockout mice have been also used to investigate the role of oxidative stress in the response to pollutants. In this process, a central role is played by Nrf2. Deficiency of Nrf2 resulted in more pronounced oxidative stress in lungs and enhanced susceptibility to ETS-induced emphysema in mice.[43,44,80] Type 2 pneumocytes from Nrf2 knockout mice showed impaired growth and increased sensitivity to oxidant-induced cell death.[81] *In vivo*, Nrf2 knockout mice showed increased susceptibility to emphysema and lung inflammation following ETS exposure, accompanied by excessive oxidative stress and increased apoptosis.[43] Moreover, wild-type mice exposed to ETS, when treated with 1-[2-cyano-3-,12-dioxooleana-1,9(11)-dien-28-oyl]imidazole (CDDO-Im), a compound increasing transcriptional induction of Nrf2-regulated antioxidant genes, exhibited significant reductions in both oxidative stress (DNA oxidation) and alveolar destruction.[82] Among inbred mouse strains, the C57BL/6J strain showed lower basal Nrf2 levels compared with strains of mice resistant to oxidative stress.[58,83]

Because of the well-demonstrated ability of Nrf2 in protecting against airway inflammation and asthma by its interaction with the antioxidant response element (ARE) and activating the phase II enzyme genes, it is possible that both Nrf2 and the phase II enzymes polymorphisms are associated with a modulation of lung diseases development as a consequence of air pollutant exposure.[84] The importance of antioxidants in protecting against CS-induced effects was also demonstrated in mice overexpressing Cu/Zn SOD. In these mice, after a 1-year exposure to CS, no neutrophilic inflammation and a reduced onset of airspace enlargement was observed.[85]

Susceptibility to O_3-induced lung inflammation and injury in mice

Significant variation between inbred strains of mice has been observed in multiple ozone toxicity phenotypes, including mortality, decreased lung function, increased pulmonary inflammation, and hyperpermeability.[59–61,86] Exposure of BALB/c mice to O_3 for 6 weeks induced a chronic inflammatory process, with alveolar enlargement and damage. Chronic exposure to O_3 upregulated proinflammatory cytokines and chemokines, such as IL-13 and IFN-γ, and increased MMP-12 expression.[87] IL-4 and IL-13 can modulate lung responses to high dose of O_3 exposure (3 ppm) as demonstrated in deficient or transgenic mice. IL-4/IL-$13^{-/-}$ and IL-$13^{-/-}$ mice developed a lesser degree of O_3-induced airway hyperresponsiveness, while IL-13–overexpressing transgenic mice developed a greater degree of damage compared with wild-type mice.[88] In C57BL/6J mice 0.5 ppm of chronic O_3 exposure increased TGF-β protein level in the epithelial lining fluid (ELF) and activated TGF-β signaling pathways, which mediates O_3-induced lung fibrotic responses *in vivo*.[89]

Prow *et al.*[90–91] used inbred mice to investigate the genetic determinants of acute lung injury. Recombinant inbred (RI) strains derived from A/J mice (more sensitive) and C57BL/6J mice (less sensitive) showed a continuous phenotypic pattern, suggesting a multigenic trait. Quantitative trait loci (QTL) and RI analyses suggested three major loci linked to O_3 susceptibility. Linkage analyses identified chromosomal segments with genes controlling susceptibility to the lung inflammatory (chromosome 17), injury (chromosome 11), and hyperpermeability (chromosome 4) responses to O_3 exposure. Interestingly, the QTL on chromosomes 11 and 17 are similar to those described by Kleeberger *et al.*[92] for susceptibility to inflammation induced by exposure to O_3. Within the chromosome 17 QTL, there are a number of candidate genes, including proinflammatory cytokines and TNF-α. Given its diverse bioregulatory activities, TNF-α may be a candidate susceptibility gene for subacute O_3-induced inflammation, as reported also for CS exposure. In fact, pretreatment with anti-TNF-α antibody attenuated O_3-induced inflammation and injury in susceptible C57BL/6J mice, but not in resistant C3H/HeJ mice.[92] After acute and subacute exposure to O_3, mice with targeted deletion of TNF receptors showed significantly reduced inflammation and epithelial cell injury relative to wild-type control mice.[35] Main downstream signal transducers of TNF-mediated lung inflammation and injury after O_3 are NF-κB and mitogen-activated protein kinase (MAPK)/AP-1.[93] Also the chromosome 11 QTL contains numerous candidate genes including *Nos2,* which encodes inducible nitric oxide synthase (iNOS) and proinflammatory chemokines and cytokines. The iNOS-deficient mice showed increased susceptibility to O_3-induced acute lung injury with significant increases in lavage protein content, MIP-2 and MMP-9 content, and polymorphonuclear leukocytes.[94] The chromosome 4 QTL contains a candidate gene, Toll-like receptor 4 (TLR4) that has been implicated in innate immunity and endotoxin susceptibility.[95–97] TLR4 has been demonstrated to have an important role in the hyperpermeability responses to O_3, as demonstrated by comparing C3H/HeOuJ and C3H/HeJ mice.[98] These strains differ only at a polymorphism in the coding region of the TLR4 gene and the polymorphism confers resistance to endotoxin-induced injury in the C3H/HeJ mouse compared to wild-type C3H/HeOuJ. Significantly greater protein concentrations and mRNA were found in C3H/HeOuJ mice compared with C3H/HeJ mice after exposure to O_3.[98] These results indicate that a QTL on chromosome 4 explains a significant portion of the genetic variance in O_3-induced hyperpermeability. Hyperoxic lung injury induces inflammation and noncardiogenic edema in the lung. Studies in mice with targeted deletion of Nrf2 confirmed a role for this gene as a potential candidate for susceptibility to hyperoxic lung injury.[99] Nrf2 has also been found important in the pathogenesis of asthma phenotypes in a mouse model.[100]

Conclusions

Unlike CS that contains high levels of pro-oxidants (such as free radicals), O_3 is not a radical species per se and its toxic effects are mediated through free radical mediators.[101] However, CS and O_3 seem to share common mechanisms inducing adverse health effects, which is corroborated by the fact that similar mouse strains respond alike to both these toxicants. CS and O_3 may elicit their effects either directly or indirectly by generation of biological active products (e.g., ROS and lipid peroxidation products). Pollutant-induced ROS or free radicals can react with lipids in the ELF or cell membrane as well as proteins and antioxidants and activate signaling pathways or secondary mediators. In addition, the recruitment of inflammatory cells into the lung contributes to tissue damage through the release of chemotactic and toxic mediators, including proteolytic enzymes, ROS, and cytokines (Fig. 1).

Mouse models have demonstrated that a few regions of the genome confer differential susceptibility to proinflammatory and ROS-inducing stimuli, suggesting that common genetic mechanisms contribute to the pathogenesis of lung injury caused by these pollutants. Because considerable homology exists between the human and the mouse genomes in these common QTL, these results may have implications for the role of these candidate genes in the pathogenesis of human pulmonary diseases

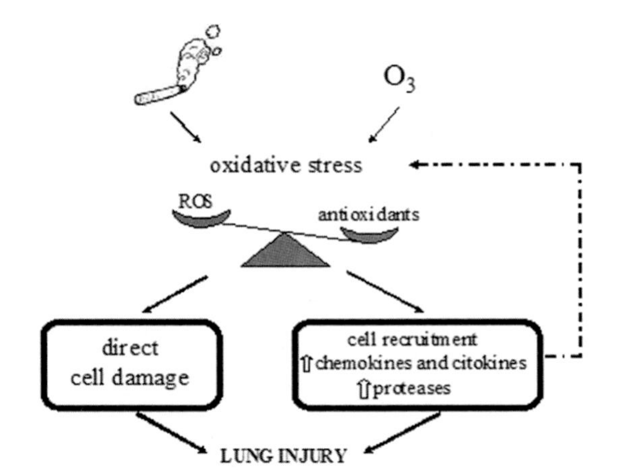

Figure 1. Common mechanisms of CS- and O_3-induced lung injury via oxidative stress. This effect can be mediated either by active biomolecules or direct cell damage.

including asthma, acute respiratory distress syndrome (ARDS), and emphysema.

In addition, the use of animal models may help to elucidate the underlying mechanisms that are behind the different susceptibility to other environmental stressors, such as diesel exhaust and ambient particulate in developing asthma. In fact, the data from clinical studies on these subjects are still ambiguous and controversial due to the high variability of the populations.[102,103]

Combined investigations across cell models, inbred mice, and humans may continue to provide important insights to understanding genetic factors that contribute to differential susceptibility to air pollutants.

Conflicts of interest

The authors declare no conflicts of interest.

References

1. Brunekreef, B. & S.T. Holgate. 2002. Air pollution and health. *Lancet* **360:** 1233–1242.
2. Budinger, G.R. & G.M. Mutlu. 2011. Update in environmental and occupational medicine 2010. *Am. J. Respir. Crit. Care Med.* **83:** 1614–1619.
3. Ko, F.W. & D.S. Hui. 2011. Air pollution and COPD. *Respirology***17:** 395–401.
4. Kelly, F.J. & J.C. Fussell. 2011. Air pollution and airway disease. *Clin. Exp. Allergy* **41:** 1059–1071.
5. Franchini, M. & P.M. Mannucci. 2012. Air pollution and cardiovascular disease. *Thromb. Res.* **129:** 230–234.
6. Ito, K. 2011. Semi-long-term mortality effects of ozone. *Am. J. Respir. Crit. Care Med.* **184:** 754–755.
7. Viegi, G. *et al.* 2006. Epidemiology of chronic obstructive pulmonary disease: health effects of air pollution. *Respirology* **11:** 523–532.
8. Nicita-Mauro, V. *et al.* 2008. Smoking, aging and the centenarians. *Exp. Gerontol.* **43:** 95–101.
9. Adcock, I.M. 2011. Chronic obstructive pulmonary disease and lung cancer: new molecular insights. *Respiration* **81:** 265–284.
10. Mudway, I.S. & F.J. Kelly. 2009. Ozone and the lung: a sensitive issue. *Mol. Aspects Med.* **21:** 1–48.
11. Kampa, M. & E. Castanas. 2008. Human health effects of air pollution. *Environ. Pollut.* **151:** 362–367.
12. Jerrett, M. *et al.* 2009. Long-term ozone exposure and mortality. *N. Engl. J. Med.* **360:** 1085–1095.
13. Bhalla, D.K. 1999. Ozone-induced lung inflammation and mucosal barrier disruption: toxicology, mechanisms, and implications. *J. Toxicol. Environ. Health B. Crit. Rev.* **2:** 31–86.
14. Castillejos, M. *et al.* 1992. Effects of ambient ozone on respiratory function and symptoms in Mexico City schoolchildren. *Am. Rev. Respir. Dis.* **145:** 276–282.
15. Church, D.F. & W.A. Pryor. 1985. Free-radical chemistry of cigarette smoke and its toxicological implications. *Environ. Health Perspect.* **64:** 111–126.
16. Pryor, W.A., D.G. Prier & D.F. Church. 1983. Electron-spin resonance study of mainstream and sidestream cigarette smoke: nature of the free radicals in gas-phase smoke and in cigarette tar. *Environ. Health Perspect.* **47:** 345–355.
17. Zang, L.Y., K. Stone & W.A. Pryor. 1995. Detection of free radicals in aqueous extracts of cigarette tar by electron spin resonance. *Free Radic. Biol. Med.* **19:** 161–167.
18. Alberg, A.J. & J.M. Samet. 2003. Epidemiology of lung cancer. *Chest* **123:** 21S–49S.
19. Sethi, J.M. & C.L. Rochester. 2000. Smoking and chronic obstructive pulmonary disease. *Clin. Chest Med.* **21:** 67–86.
20. Mannino, D.M. & A.S. Buist. 2007. Global burden of COPD: risk factors, prevalence, and future trends. *Lancet* **370:** 765–773.
21. Department of Health and Human Services. 2004. *The health consequences of smoking. A Report of the Surgeon General.* U.S. Department of Health and Human Services, Washington, DC.
22. Allender S. *et al.* 2008. European cardiovascular disease statistics. www.ehnheart.org/projects/ehhi
23. Brennan, P. *et al.* 2006. High cumulative risk of lung cancer death among smokers and nonsmokers in Central and Eastern Europe. *Am. J. Epidemiol.* **164:** 1233–1241.
24. Bhalla, D.K. 2002. Interactive effects of cigarette Smoke and ozone in the induction of lung injury. *Toxicol. Sci.* **65:** 1–3.
25. Churg, A. 2003. Interactions of exogenous or evoked agents and particles: the role of reactive oxygen species. *Free Radic. Biol. Med.* **34:** 1230–1235.
26. Yu, M., K.E. Pinkerton & H. Witschi. 2002. Short-term exposure to aged and diluted sidestream cigarette smoke enhances ozone-induced lung injury in B6C3F1 mice. *Toxicol. Sci.* **65:** 99–106.
27. March, T.H. *et al.* 2002. Effects of concurrent ozone exposure on the pathogenesis of cigarette smoke-induced emphysema in B6C3F1 mice. *Inhal. Toxicol.* **14:** 1187–1213.
28. Pryor, W.A., G.L. Squadrito & M. Friedman. 1995. The cascade mechanism to explain ozone toxicity: the role of

lipid ozonation products. *Free Radic. Biol. Med.* **19:** 935–941.

29. Valacchi, G. *et al.* 2004. In vivo ozone exposure induces antioxidant/stress-related responses in murine lung and skin. *Free Radic. Biol. Med.* **36:** 673–681.
30. Heijink, I.H. *et al.* 2012. Cigarette smoke impairs airway epithelial barrier function and cell-cell contact recovery. *Eur. Respir. J.* **39:** 419–428.
31. Ciencewicki, J., S. Trivedi & S.R. Kleeberger. 2008. Oxidants and the pathogenesis of lung diseases. *J. Allergy Clin. Immunol.* **122:** 456–468.
32. Murugan, V. & M.J. Peck. 2009. Signal transduction pathways linking the activation of alveolar macrophages with the recruitment of neutrophils to lungs in chronic obstructive pulmonary disease. *Exp. Lung Res.* **35:** 439–485.
33. Song, H., W. Tan & X. Zhang. 2011. Ozone induces inflammation in bronchial epithelial cells. *J. Asthma* **48:** 79–83.
34. Bhalla, D.K. *et al.* 2009. Cigarette smoke, inflammation, and lung injury: a mechanistic perspective. *J. Toxicol. Environ. Health* **12:** 45–64.
35. Cho, H.Y., L.Y. Zhang & S.R. Kleeberger. 2001. Ozone-induced lung inflammation and hyperreactivity are mediated via tumor necrosis factor alpha receptors. *Am. J. Physiol. Lung Cell. Mol. Physiol.* **280:** L537–L546.
36. Yao, H. & I. Rahman. 2011. Current concepts on oxidative/carbonyl stress, inflammation and epigenetics in pathogenesis of chronic obstructive pulmonary disease. *Toxicol. Appl. Pharmacol.* **254:** 72–85.
37. Chung, K.F. & I.M. Adcock. 2008. Multifaceted mechanisms in COPD: inflammation, immunity, and tissue repair and destruction. *Eur. Respir. J.* **31:** 1334–1356.
38. Hicks, A. *et al.* 2010. Effects of LTB4 receptor antagonism on pulmonary inflammation in rodents an non-human primates. *Prostaglandins Other Lipid Mediat.* **92:** 33–43.
39. Mudway, I.S. *et al.* 1999. Compromised concentrations of ascorbate in fluid lining the respiratory tract in human subjects after exposure to ozone. *Occup. Environ. Med.* **56:** 473–481.
40. Avissar, N.E. *et al.* 2000. Ozone, but not nitrogen dioxide, exposure decreases glutathione peroxidases in epithelial lining fluid of human lung. *Am. J. Respir. Crit. Care Med.* **162:** 1342–1347.
41. Rahman, I. & W. MacNee. 1999. Lung glutathione and oxidative stress: implications in cigarette smoke-induced airway disease. *Am. J. Physiol. Lung Cell. Mol. Physiol.* **277:** L1067–L1088.
42. Rahman, I. & I.M. Adcock. 2006. Oxidative stress and redox regulation of lung inflammation in COPD. *Eur. Respir.* **28:** 219–242.
43. Cho, H.Y. & S.R. Kleeberger. 2010. Nrf2 protects against airway disorders. *Toxicol. Appl. Pharmacol.* **244:** 43–56.
44. Rangasamy, T. *et al.* 2004. Genetic ablation of Nrf2 enhances susceptibility to cigarette smoke-induced emphysema in mice. *J. Clin. Invest.* **114:** 1248–1259.
45. Iizuka, T. *et al.* 2005. Nrf2-deficient mice are highly susceptible to cigarette smoke induced emphysema. *Genes Cells.* **10:** 1113–1125.
46. Cho, H.Y. *et al.* 2005. Gene expression profiling of Nrf2-mediated protection against oxidative injury. *Free Radic. Biol. Med.* **38:** 325–343.
47. Annesi-Maesano, I. *et al.* 2003. Subpopulations at increased risk of adverse health outcomes from air pollution. *Eur. Respir. J.* **40:** 57–63.
48. Gauderman, W.J. *et al.* 2004. The effect of air pollution on lung development from 10 to 18 years of age. *N. Engl. J. Med.* **351:** 1057–1067.
49. Holz, O. *et al.* 1999. Ozone-induced airway inflammatory changes differ between individuals and are reproducible. *Am. J. Respir. Crit. Care Med.* **159:** 776–784.
50. Fletcher, C. 1977. The natural history of chronic airflow obstruction. *Br. Med. J.* **1:** 1645–1648.
51. Salvi, S.S. & P.J. Barnes. 2009. Chronic obstructive pulmonary disease in non-smokers. *Lancet.* **374:** 733–743.
52. Valacchi, G. *et al.* 2007. Lung vitamin E transport processes are affected by both age and environmental oxidants in mice. *Toxicol. Appl. Pharmacol.* **222:** 227–234.
53. Sullivan, J.B. & G.R. Krieger, eds. 2001. *Clinical Environmental Health and Toxic Exposure*. Lippincott Williams and Wilkins, Philadelphia, PA.
54. Kleeberger, S.R. 2003. Genetic aspects of susceptibility to air pollution. *Eur. Respir. J.* **40:** 52–56.
55. Kleeberger, S.R. 2005. Genetic aspects of pulmonary responses to inhaled pollutants. *Exp. Toxicol. Pathol.* **57:** 147–153.
56. Yao, H. *et al.* 2008. Cigarette smoke-mediated inflammatory and oxidative responses are strain-dependent in mice. *Am. J. Physiol. Lung Cell. Mol. Physiol.* **294:** L1174–1186.
57. Guerassimov, A. 2004. The development of emphysema in cigarette smoke-exposed mice is strain dependent. *Am. J. Respir. Crit. Care Med.* **170:** 974–980.
58. Vecchio, D. *et al.* 2010. Reactivity of mouse alveolar macrophages to cigarette smoke is strain dependent. *Am. J. Physiol. Lung Cell. Mol. Physiol.* **298:** L704–L713.
59. Bauer, A.K. & S.R. Kleeberger. 2010. Genetic mechanisms of susceptibility to ozone-induced lung disease. *Ann. N.Y. Acad. Sci.* **1203:** 113–119.
60. Cho, H.Y. & S.R. Kleeberger. 2007. Genetic mechanisms of susceptibility to oxidative lung injury in mice. *Free Radic. Biol. Med.* **42:** 433–445.
61. Savov, J.D. *et al.* 2004. Ozone-induced acute pulmonary injury in inbred mouse strains. *Am. J. Respir. Cell Mol. Biol.* **31:** 69–77.
62. Young, R.P. *et al.* 2006. Functional variants of antioxidant genes in smokers with COPD and in those with normal lung function. *Thorax.* **61:** 394–399.
63. Sandford, A.J. *et al.* 2001. Susceptibility genes for rapid decline of lung function in the lung health study. *Am. J. Respir. Crit. Care Med.* **163:** 469–473.
64. Kucukaycan, M. *et al.* 2002. Tumor necrosis factor-alpha +489 G/A gene polymorphism is associated with chronic obstructive pulmonary disease. *Respir. Res.* **3:** 29–35.
65. Celedon, J.C. *et al.* 2004. The transforming growth factor-beta1 (TGFb1)gene is associated with chronic obstructive pulmonary disease (COPD). *Hum. Mol. Genet.* **13:** 1649–1656.
66. Bartalesi, B. *et al.* 2005. Different lung responses to cigarette smoke in two strains of mice sensitive to oxidants. *Eur. Respir. J.* **25:** 15–22.

67. Cavarra, E. *et al.* 2001. Effects of cigarette smoke in mice with different levels of α_1-proteinase inhibitor and sensitivity to oxidants. *Am. J. Respir. Crit. Care Med.* **164:** 886–890.
68. Martorana, P.A. *et al.* 2006. Models for COPD involving cigarette smoke. *Drug Discov. Today Dis. Models* **3:** 225–230.
69. March, T.H. *et al.* 1999. Cigarette smoke exposure produces more evidence of emphysema in B6C3F1 mice than in F344 rats. *Toxicol. Sci.* **51:** 289–299.
70. Hautamaki, R.D. *et al.* 1997. Requirement for macrophage elastase for cigarette smoke-induced emphysema in mice. *Science* **277:** 2002–2004.
71. Cavarra, E. *et al.* 2001. Human SLPI inactivation after cigarette smoke exposure in a new in vivo model of pulmonary oxidative stress. *Am. J. Physiol. Lung Cell. Mol. Physiol.* **281:** L412–L417.
72. Rahman, I. 2012. Pharmacological antioxidant strategies as therapeutic interventions for COPD. *Biochim. Biophys. Acta* **1822:** 714–728.
73. Morris, D.G. *et al.* 2003. Loss of integrin avb6-mediated TGF-b activation causes MMP12-dependent emphysema. *Nature* **422:** 169–173.
74. Shapiro, S.D. 1994. Elastolytic metalloproteinases produced by human mononuclear phagocytes. Potential roles in destructive lung disease. *Am. J. Respir. Crit. Care Med.* **150:** S160–S164.
75. Hoshino, T. *et al.* 2007. Pulmonary inflammation and emphysema: role of the cytokines IL-18 and IL-13. *Am. J. Respir. Crit. Care Med.* **176:** 49–62.
76. Zheng, T. *et al.* 2000. Inducible targeting of IL-13 to the adult lung causes matrix metalloproteinase- and cathepsin-dependent emphysema. *J. Clin. Invest.* **106:** 1081–1093.
77. Wang, Z. *et al.* 2000. Interferon-g induction of pulmonary emphysema in the adult murine lung. *J. Exp. Med.* **192:** 1587–1600.
78. Kang, M.J. *et al.* 2007. IL-18 is induced and IL-18 receptor a plays a critical role in the pathogenesis of cigarette smoke-induced pulmonary emphysema and inflammation. *J. Immunol.* **178:** 1948–1959.
79. Churg, A. *et al.* 2004. Tumor necrosis factor-a drives 70% of cigarette smoke induced emphysema in the mouse. *Am. J. Respir. Crit. Care Med.* **170:** 492–498.
80. Rangasamy, T. *et al.* 2009. Cigarette smoke-induced emphysema in A/J mice is associated with pulmonary oxidative stress, apoptosis of lung cells, and global alterations in gene expression. *Am. J. Physiol. Lung Cell. Mol. Physiol.* **296:** L888–L900.
81. Reddy, N.M. *et al.* 2007. Deficiency in Nrf2–GSH signaling impairs type II cell growth and enhances sensitivity to oxidants. *Am. J. Respir. Cell. Mol. Biol.* **37:** 3–8.
82. Sussan, T.E. *et al.* 2009. Targeting Nrf2 with the triterpenoid CDDO-imidazolide attenuates cigarette smoke-induced emphysema and cardiac dysfunction in mice. *Proc. Natl. Acad. Sci. USA* **106:** 250–255.
83. Cho, H.Y. *et al.* 2002. Linkage analysis of susceptibility to hyperoxia. *Am. J. Respir. Cell Mol. Biol.* **26:** C42–C51.
84. Li, N. & A.E Nel. 2006. Role of the Nrf2-mediated signaling pathway as a negative regulator of inflammation: implications for the impact of particulate pollutants on asthma. *Antioxid. Redox Signal.* **8:** 88–98.
85. Foronjy, R.F. *et al.* 2006. Superoxide dismutase expression attenuates cigarette smoke- or elastase-generated emphysema in mice. *Am. J. Respir. Crit. Care Med.* **173:** 623–631.
86. Wesselkamper, S.C. *et al.* 2000. Genetic susceptibility to irritant-induced acute lung injury in mice. *Am. J. Physiol. Lung Cell. Mol. Physiol.* **279:** L575–L582.
87. Triantaphyllopoulos, K. *et al.* 2011. A model of chronic inflammation and pulmonary emphysema after multiple ozone exposures in mice. *Am. J. Physiol. Lung Cell Mol. Physiol.* **300:** L691–L700.
88. Williams, A.S. *et al.* 2008. Modulation of ozone-induced airway hyperresponsiveness and inflammation by interleukin-13. *Eur. Respir. J.* **32:** 571–578.
89. Katre, A. *et al.* 2011. Increased transforming growth factor beta 1 expression mediates ozone-induced airway fibrosis in mice. *Inhal. Toxicol.* **23:** 486–494.
90. Prows, D.R. *et al.* 1997. Genetic analysis of ozone-induced acute lung injury in sensitive and resistant strains of mice. *Nat. Genet.* **17:** 471–474.
91. Prows, D.R. *et al.* 1999. Ozone-induced acute lung injury:genetic analysis of F2 mice generated from A/J and C57BL/6J strains. *Am. J. Physiol.* **277:** L372–L380.
92. Kleeberger, S.R. *et al.* 1997. Linkage analysis of susceptibility to ozone-induced lung inflammation in inbred mice. *Nat. Genet.* **17:** 475–478.
93. Cho, H.Y. *et al.* 2007. Signal transduction pathways of tumor necrosis factor-mediated lung injury induced by ozone. *Am. J. Respir. Crit. Med.* **15:** 829–839.
94. Nicholas, J. *et al.* 2002. Susceptibility to ozone-induced acute lung injury in iNOS-deficient mice. *Am. J. Physiol. Lung Cell. Mol. Physiol.* **282:** L540–L545.
95. Morris, G.E. *et al.* 2006. Cooperative molecular and cellular networks regulate Toll-like receptor-dependent inflammatory responses. *FASEB J.* **20:** 2153–2155.
96. Poltorak, A. *et al.* 1998. Defective LPS signalling in C3H/HeJ and C57BL/10ScCr mice: mutations in the *Tlr4* gene. *Science* **282:** 2085–2088.
97. Qureshi, S.T. *et al.* 1999. Endotoxin intolerant mice have mutations in toll-like receptor 4 (*Tlr4*). *J. Exp. Med.* **189:** 615–625.
98. Kleeberger, S.R. *et al.* 2000. Genetic susceptibility to ozone-induced lung hyperpermeability. Role of toll-like receptor 4. *Am. J. Respir. Cell Mol. Biol.* **22:** 620–627.
99. Cho, H.Y. *et al.* 2002. Role of NRF2 in protection against hyperoxic lung injury in mice. *Am. J. Respir. Cell Mol. Biol.* **26:** 175–182.
100. Rangasamy, T. *et al.* 2005. Disruption of Nrf2 enhances susceptibility to severe airway inflammation and asthma in mice. *J. Exp. Med.* **202:** 47–59.
101. Sticozzi, C. & G. Valacchi. 2011. Troposphere ozone as a source of oxidative stress in cutaneous tissues. *J. Sci. Ind. Res.* **70:** 918–922.
102. McCreanor, J. *et al.* 2007. Respiratory effects of exposure to diesel traffic in persons with asthma. *N. Engl. J. Med.* **357:** 2348–58.
103. Adewole, F. *et al.* 2009. Diesel exhaust causing low-dose irritant asthma with latency? *Occup. Med. (Lond).* **59:** 424–427.

Ann. N.Y. Acad. Sci. ISSN 0077-8923

ANNALS OF THE NEW YORK ACADEMY OF SCIENCES
Issue: *Environmental Stressors in Biology and Medicine*

Age-related changes in cellular protection, purification, and inflammation-related gene expression: role of dietary phytonutrients

Angela Mastaloudis and Steven M. Wood
Nu Skin Enterprises, Provo, Utah

Address for correspondence: Angela Mastaloudis, Nu Skin Enterprises, Center for Anti-Aging Research, 75 West Center Street, Provo, UT 84601. amastaloudis@nuskin.com

Oxidative injury and inflammation are intimately involved in the aging process and the development of age-related diseases. To date, most nutritional antiaging strategies have focused solely on the delivery of exogenous antioxidants to combat the negative effects of aging. A promising new strategy is to identify nutrients and phytochemicals that can directly target intrinsic cytoprotective mechanisms, including modulation of the expression of (1) genes involved in the detoxification of xenobiotics, (2) genes involved in the synthesis and regulation of intrinsic antioxidants and antioxidant enzymes, (3) genes involved in the regulation of inflammation, and (4) vitagenes. The purpose of this review is to provide an overview of the age-related changes in gene expression related to oxidative stress, detoxification, and inflammatory processes, and to discuss natural compounds with the potential to oppose age-related changes in gene expression related to these processes, which therefore may be suitable for use in human antiaging research.

Keywords: broccoli; detoxification; grape seed extract; blood orange; phase II enzymes; antioxidants

Introduction

Aging is a complex, multifactorial process that remains poorly understood. However, it is widely accepted that oxidative injury and inflammation are intimately involved in the aging process and the development of age-related diseases. One hallmark of the aging process is the dysregulation of gene expression associated with antioxidant status, cellular detoxification, and inflammatory pathways, and the accumulation of damaging metabolic by-products. Damage caused by environmental and biochemical toxicants is believed to be a major source of age-related dysregulation of gene expression. Examples of toxicants include reactive oxygen species (ROS), reactive nitrogen species, chemical carcinogens, advanced glycation end products (AGEs), and environmental pollutants. To date, most dietary antiaging strategies have focused solely on the delivery of exogenous antioxidants to boost antioxidant status in an effort to protect against toxicant-induced oxidative and inflammatory stress as a means to prevent or combat the negative effects of aging. A promising new strategy intends to identify nutrients and phytochemicals with the ability to directly target and enhance intrinsic cytoprotective mechanisms, including modulation of the expression of genes involved in the detoxification of xenobiotics and their metabolites, genes involved in the synthesis and regulation of intrinsic antioxidants and antioxidant enzymes, genes involved in the regulation of inflammation, and vitagenes. In this way, these phytonutrients may prevent damage by inducing multiple cytoprotective pathways—inducing vitagenes, upregulating antioxidant protection and anti-inflammatory activity, intercepting and detoxifying damaging compounds, and efficiently removing said toxicants before they can initiate further damage.

Cytoprotective mechanisms

Humans and other mammals possess a multitude of cytoprotective mechanisms against environmental and biochemical damage. Extrinsic antioxidants

doi: 10.1111/j.1749-6632.2012.06610.x

include diet-derived nutrients, such as vitamin E, vitamin C, and carotenoids. There are also hundreds of plant polyphenolic compounds with antioxidant capability, including tannins, flavonoids, hydroxybenzoic acids, hydroxycinnamic acids, stilbenes, and lignans.[1] While these polyphenols do have antioxidant capabilities *in vitro*, low bioavailability and low tissue concentrations make it unlikely that they act directly as antioxidants *in vivo*.[2] Instead, it is more likely that polyphenols act as indirect antioxidants, inducing the production of intrinsic antioxidants and antioxidant enzymes via nuclear factor erythroid 2–related factor 2 (Nrf2). Nrf2 is the transcription factor that is responsible for both inducible and constitutive expression of antioxidant response element (ARE)–regulated genes, including those coding for a number of antioxidant proteins and phase II detoxifying enzymes that defend against electrophilic stressors and oxidative insults.[3,4] These include the antioxidant glutathione (GSH); antioxidant enzymes, such as superoxide dismutase (SOD); and detoxification enzymes, including glutathione S-transferase 1 (GSTA1). Another cytoprotective category are the vitagenes, defined by Calabrese *et al.*[5] as "a group of genes strictly involved in preserving cellular homeostasis during stressful conditions." Examples of vitagenes include genes that code for heme oxygenase-1, heat shock proteins, thioredoxin, and sirtuin proteins.[5] One mechanism through which cytoprotective pathways are induced is hormesis. Hormesis can be described as an event in which exposure to low levels of a stressor, such as a toxicant, elicits an adaptive response such that the organism is protected against subsequent exposure to higher levels of the same stressor, levels that would otherwise be toxic.[5–7] Cytoprotective pathways, including Nrf2 and vitagenes, are impaired with aging and ultimately controlled by age-related changes in gene expression. Studies in aged rats have reported as much as a ~50% reduction in nuclear Nfr2 levels and ARE binding in liver.[3,8] Loss of Nrf2 with age is reversible and therefore an attractive target for antiaging interventions. For example, lipoic acid administration restored nuclear Nrf2 to youthful levels within 24 h of administration; levels had returned to baseline by 48 h, indicating that repeated intake is necessary to maintain effects.[8]

A number of Nrf2-regulated genes and gene products also change in expression and activity with age. GSH is a ubiquitous water-soluble, intrinsic antioxidant that exists in both the sulfhydryl (reduced GSH) and disulfide (oxidized glutathione (GSSG)) forms. *De novo* GSH synthesis, GSH recycling, and the GSH/GSSG ratio are all impaired with aging,[9] creating a more pro-oxidant environment concomitant with age-related increases in oxidative damage and toxicant load. Glutamate-cysteine ligase (GCL) is the first and rate-limiting enzyme of GSH synthesis and overexpression of GCL in fruit flies extends life span.[10] Suh *et al.*[8] demonstrated that total GSH levels were depleted by nearly 50% in rat liver with age; this decline corresponded with declines in GCL activity and nuclear Nrf2 levels. Impaired GSH synthesis has also been linked to reduced expression of glutathione synthase, the enzyme required for the final synthesis step.[11,12]

Impairments in GSH-related enzymes and GSH recovery pathways with age have also been observed, the activities of which are outlined in Figure 1. Glutathione reductase (GR), a central enzyme in cellular antioxidant defense, plays a critical role in the maintenance of the GSH/GSSG ratio by reducing oxidized GSSG back to the sulfhydryl (GSH) form.[7] Both GR expression and activity are impaired with aging in the rat liver, and higher expression is inversely associated with markers of oxidative stress.[3] GSTA1 functions in the detoxification of electrophilic compounds, including carcinogens, environmental toxins, and products of oxidative stress, by irreversible conjugation with GSH, depleting GSH in the process.[12] Transcription and activity of GSTA1 is one of the strongest correlates of longevity among the detoxification enzymes across species.[4,13] NADH:quinone oxidoreductase 1 (NQO1) is an enzyme involved in the maintenance of cell membrane integrity through regeneration of coenzyme Q_{10}[14] and is considered a reliable surrogate marker of Nrf2 activation.[15] NQO1 expression and activity are impaired with aging in the liver,[3] and higher expression has been observed in long-lived animal models.[14,16] The vitagene HO1 codes for the inducible isoform of heme oxygenase-1 (HO-1), the rate-limiting enzyme of heme degradation; it has potent antioxidant and anti-inflammatory functions. Higher expression of HO-1 in heart and liver, but not brain, of long-lived models compared to controls has been reported.[16] Glutathione peroxidase (GPx) is an endogenous antioxidant enzyme involved in the catalytic removal of hydrogen

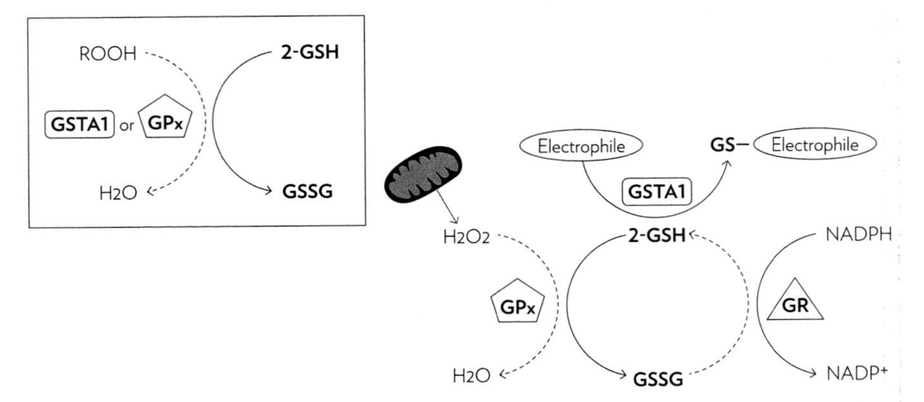

Figure 1. Outline of the antioxidant and detoxification mechanisms of GSH and GSH-related enzymes. Detoxification activity of GSH:GSTA1 catalyzes the detoxification of electrophilic compounds by irreversible conjugation with GSH, depleting GSH in the process. Direct antioxidant activity of GSH:GPx and GSTA1 are responsible for catalyzing the reduction of reactive compounds, for example, lipid hydroperoxides and H_2O_2 by GSH in cytosol and in the mitochondria. These enzymes play a critical protective role in the mitochondria eliminating H_2O_2, as the mitochondria does not contain catalase. GSH maintenance: GR maintains intracellular GSH levels and the GSH/GSSG ratio by catalyzing the reduction of GSSG back to GSH; NAD(P)H is a cofactor in this reaction. Abbreviations: GSH, glutathione; GSSG, glutathione disulfide; GR, glutathione reductase; GSTA1, glutathione S-transferase 1; GPx, glutathione peroxidase.

peroxide (H_2O_2) in a reaction that requires GSH as an electron donor.[7] GPx activity is decreased with aging, potentially exacerbating mitochondrial oxidative damage.[14] SOD and catalase (CAT) are endogenous antioxidant enzymes involved in the catalytic removal of superoxide ($O_2^{\bullet -}$) and H_2O_2, respectively.[7] The expression of each is dysregulated with age, especially in the presence of additional stressors, and likely contributes to the accumulation of oxidative damage observed with aging.[7,17] Trx is the vitagene that codes for the thioredoxin (Trx) protein, an intrinsic antioxidant enzyme. Trx is involved in the repair of protein damage and is responsible for the reduction of oxidized thiols, it is especially active in neuronal cells.[7,18] Transgenic mice overexpressing Trx are more resistant to ischemia-reperfusion injury in the brain than wild types,[18] and higher expression in heart, liver, and brain of long-lived models compared to controls has been observed.[16] While this is not an exhaustive list of Nrf2-regulated genes, it is representative of the dysregulation of expression of Nrf2-regulated genes with aging, and in some cases, optimal expression related to longevity.

The expression of many inflammation-related genes is similarly dysregulated with age. In fact, systemic chronic inflammation increases with age, even in the absence of risk factors or chronic disease.[19] The transcription factor NF-κB is a master regulator of inflammation and is consistently upregulated with aging.[16] Greater expression of NF-κB is associated with higher levels of inflammatory biomarkers in humans, including C-reactive protein (CRP) and proinflammatory cytokines (e.g., IL-6 and TNF-α).[19] Inducible nitric oxide synthase (iNOS) is responsible for the production of nitric oxide, a free radical, which acts as a biologic mediator in several processes, including neurotransmission. Expression of iNOS is dramatically increased with age, even in healthy animals in the absence of extrinsic stressors.[20] The activation of iNOS with age can be prevented with caloric restriction, an established anti-aging intervention.[20] Endothelium-derived NOS (eNOS) is responsible for the production of nitric oxide necessary for vasodilation. Dysregulated expression of eNOS with aging has been associated with cardiovascular risk factors, including impairments in vasodilation and blood flow.[21] Cyclooxygenase 2 (COX2) is the key enzyme in prostaglandin biosynthesis; dysregulated expression of COX2 and associated inflammation have been observed in a number of aging models.[21–23]

Role of dietary phytochemicals

Age-related changes in gene expression lead to a loss of cellular homeostasis characterized by a pro-oxidative state, contributing to the accumulation of toxicants and metabolic damage products. One approach to oppose age-related dysregulation of gene expression is to intervene early in life, even in midlife, with nutritional compounds that influence the expression of genes, including vitagenes, in key pathways to protect against and/or remove cellular toxicants and enhance repair mechanisms. Such interventions may prevent, delay, and/or attenuate the aging process. The intervention dosing is especially critical in the case that these phytochemicals are cytoprotective via hormetic effects, where low doses elicit an adaptive response, but high levels may be toxic.[5,6] While phytochemicals have been well studied for their ability to prevent or treat chronic conditions, their antiaging capabilities have been little studied, and most often doses have not been well controlled. By examining their effects on expression of genes known to change with aging, it may be possible to identify ingredients that could be protective against the negative effects of aging. Below is a brief overview of select ingredients that have been shown to modulate the expression of genes related to the Nrf2/phase II detoxification and inflammatory pathways in a number of oxidative stress and disease conditions associated with premature aging.

Broccoli seed extract

The health benefits of cruciferous vegetables have been attributed to their high levels of glucosinolates, glucoraphinin (GRP) being the most abundant form found in broccoli.[24] GRP is metabolized to the isothiocyanate sulforaphane (SFE) by the enzyme myrosinase. Myrosinase is found both within the plant wall of cruciferous vegetables and in the human gut. The highest levels of glucosinolates are found in broccoli seeds and sprouts, and therefore, recent research has focused on the health benefits of SFE derived from seeds and sprouts. To date, SFE is considered the most potent naturally occurring inducer of phase II detoxification enzymes via the Nrf2 transcription factor pathway.[25] Using Nrf2 wild-type and knockout (KO) mice, Thimmulappa *et al.*[26] identified genes induced by SFE in small intestine using microarray analysis. Of the Nrf2-regulated genes, GSTs, GCL, carboxyl esterase, and NQO1 were the most potently induced by SFE. In a similar study design, Hu *et al.*[27] used Nrf2 KO mice to identify Nrf2-dependent genes that are inducible by SFE in liver including GCL, GSTAs, HO-1, Trx, GPx, GR, and epoxide hydrolase. These studies reveal the primary genes through which SFE is likely to exert its benefits, including its cardioprotective effects. For example, Xue *et al.*[28] reported inductions of GR and NQO1 in human endothelial cells, paralleled by increased nuclear translocation of Nrf2 and attenuation of intracellular ROS formation under both normoglycemic and hyperglycemic (prooxidative) conditions. Similar cardioprotective effects were observed in rat cardiomyocytes where expression and activity of GR, GSTA1, Trx, and NQO1, but not GPx, were induced by SFE. Corresponding increases in GSH concentration, enhanced cell viability, and a decrease in ROS production were also observed.[29] SFE was also cytoprotective against both H_2O_2 and paraquat-generated cytotoxicity in neuronal cells, inducing expression of GCL and HO-1 and increasing levels of GSH and enhancing cell viability.[30] SFE was similarly protective in an animal model of traumatic brain injury (TBI). Administration of SFE by i.p. 15 min post-TBI was protective against neurological function deficits, neuronal cell death, and oxidative damage resulting from the brain injury. Furthermore, expression of Nrf2, HO-1, and NQO-1 were significantly increased in the cortex of treated animals 24 h post-TBI,[31] indicating that the protection was conferred via induction of the Nrf2/ARE pathway.

Unlike conventional antioxidants, SFE treatment has been demonstrated to protect against exercise-induced muscle damage. In rats subjected to exhaustive exercise, SFE feeding blunted markers of muscle damage (creatine kinase and lactate dehydrogenase) and lipid peroxidation as well as prevented myofibular damage assessed by histology. The positive histology data were paralleled by enhanced protein expression and activity of Nrf2, GST, GR, and NQO1, but not GPx or Trx in skeletal muscle.[32] In contrast, Trx was activated by SFE from broccoli in a mouse model of myocardial infarction (MI); expression of HO-1 and Nrf2 were also upregulated. The hearts from the broccoli-supplemented group had greater left ventricular function pre-MI, a reduced MI size, and an attenuation of MI-induced cardiomyocyte apoptosis.[33] The differential effects of SFE on Trx in skeletal muscle and heart muscle highlight the need to examine effects directly in the tissue of interest.

While there is ample epidemiological evidence for the health benefits of cruciferous vegetables,[34] there are limited human clinical data available regarding the role of broccoli products in detoxification, antioxidant protection, and inflammatory balance. Gasper *et al.*[35] reported that a single serving of GRP-rich broccoli, but not conventional broccoli, upregulated Trx, GCL, and aldoketoreductases in the gastric mucosa of human subjects. These data confirm that broccoli extracts containing GRP can induce phase II detoxification enzymes in humans and in a short time frame (6 h). Similarly, gene expression analysis in human nasal lavage cells revealed a dose-dependent induction of GSTM1, NQO1, HO-1, and GST following supplementation with SFE-rich broccoli seed extracts over three days; doses ranged from 25 to 200 g/day.[36] The use of a Loci-I-Gut perfusion tube allowed for the sampling of human enterocytes pre- and postexposure to a broccoli and onion extract. GSTA1 and UDP-glucuronosyltransferase (UGT1A1) were both induced in response to the extract, but follow-up studies in Caco-2 cells suggest that GSTA1 was induced by the broccoli and UGT1A1 by the onion extract.[37] By contrast, four weeks of supplementation with a sprout blend that included broccoli sprouts, but did not account for SFE or GRP content, had no effect on the expression of any of the phase II enzymes assessed in red blood cells. The blend did confer some benefit—lymphocytes collected from the subjects were more resistant to H_2O_2-induced DNA damage compared to controls.[38] Explanations for this unexpected lack of effect on phase II enzymes could be the length of the treatment, the absence of SFE or GRP in the sprout blend, and/or the nature of the tissue studied. These findings highlight the importance of using standardized ingredients and measuring gene expression in the tissue of interest.

Smoking is a major source of toxicant exposure, oxidative stress, and a proinflammartory state. Consumption of 100 g of fresh broccoli sprouts/day for one week in smokers led to decreases in total and LDL cholesterol in men and an increase in HDL cholesterol in women; markers of lipid peroxidation, including 8-isoprostanes, decreased significantly in both males and females in the treatment group. Phase II enzymes were not measured, but the reduced to oxidized CoQ_{10} ratio ($CoQ_{10}H_2/CoQ_{10}$) increased significantly in response to the treatment, which may be attributed to an increased expression of NQO1, lending further evidence for the efficacy of SFE in humans.[39]

The chemopreventative, cardiovascular, neuroprotective, and anti-inflammatory actions of broccoli seeds and sprouts make them unique and potentially potent antiaging ingredients meriting further study in humans.

Grape seed extract

Humans have consumed grapes (including the skin and seeds) and grape products (e.g., wine, raisins, juice, etc.) for centuries, and dietary intakes are associated with a multitude of health benefits, including maintenance of healthy blood pressure, cardiovascular health, anti-inflammatory benefits, blood glucose modulation, and exercise performance. These health benefits have been attributed to their high content of polyphenolic compounds including proanthocyanidins (PACs). Grape seed extract (GSE) contains the highest concentrations of grape polyphenolics, and standardized grape seed extracts have been developed to deliver known amounts of PACs. One of the primary mechanisms of GSE is as an anti-inflammatory agent; evidence suggests that GSE has both targeted and systemic anti-inflammatory actions. For example, GSE added to high-fat proinflammatory diets for 19 weeks prevented the upregulation of proinflammatory cytokines CRP, TNF-α, and IL-6 in visceral fat compared to the control group.[40,41] At the same time, GSE supplementation upregulated the expression of the anti-inflammatory cytokine adiponectin in visceral fat. TNF-α and CRP were also attenuated in the plasma, overall indicating a systemic reduction of inflammation with GSE fortification. Interestingly, GSE prevented weight gain despite the *ad libitum* feeding of the high-fat diets.[40,41]

The brain is one type of tissue in which GSE has been demonstrated to exert targeted anti-inflammatory benefits, as evidenced by research in a transgenic mouse model of Alzheimer's disease (AD). Inclusion of GSE in the diet for nine months had dramatic effects on inflammation and disease pathology, eliciting significant reductions in several parameters of AD pathology, including misfolded amyloid protein (AB) concentrations in the plasma, AB deposition in the neocortex and hippocampus, and total AB plaque burden in the brain. Significantly lower levels of micro-, but not astrogliosis, were also observed with GSE fortification.[42]

Several human clinical trials have also demonstrated benefits of GSE against oxidative and inflammatory stressors associated with chronic conditions. For example, supplementation of middle-aged subjects with moderately high cholesterol at a dose of 200 or 400 mg/day of GSE for 12 weeks led to reductions in lipid peroxidation in both treatment groups.[43] Similarly, when subjects with metabolic syndrome were supplemented with either 150 or 300 mg GSE/day for four weeks, systolic and diastolic blood pressure was significantly lowered in both treatment groups, suggesting a cardioprotective vasodilatory effect of GSE, potentially via eNOS.[44] Chronic hyperglycemia and diabetes are associated with accelerated aging characterized by oxidative damage and inflammation and an accumulation of toxicants including AGEs. Treatment of rats with GSE before and for three days following exposure to a diabetes-inducing chemical, alloxan, protected pancreatic tissue from oxidative damage. Pancreatic tissues of GSE-treated animals had lower levels of damaging nitrites/nitrates and higher levels of reduced GSH 48- and 72-h post-alloxan exposure compared to the control group. Furthermore, GSE attenuated hyperglycemia and modulated serum insulin levels.[45] Diabetic patients supplemented with 600 mg GSE/day for four weeks experienced increased levels of reduced GSH and lower levels of CRP.[46] The increased levels of reduced GSH suggest a role for Nrf2 in the protective effect of GSE against the oxidative stress of diabetes in humans.

A conclusion, based on its anti-inflammatory and antioxidant activity, is that GSE also has the potential to attenuate the damage caused by environmental toxicants. Chronic heavy smokers supplemented with 150 mg of GSE/day for four weeks showed improvements in antioxidant status and increased resistance to lipid peroxidation suggesting cardioprotective benefits of GSE in a high-risk group.[47]

Taken together, GSE appears to elicit protective effects against oxidative damage and inflammation in a host of conditions, and it does so even within a short duration of treatment. Further research on the antiaging effects of GSE is merited.

Blood (red) orange

Oranges (*Citrus sinensis*) are another food source rich in phytochemicals with antiaging, cytoprotective benefits. Some of the primary bioactive ingredients in oranges are vitamin C, flavanones (e.g., naringenin, hesperetin), β-cryptoxanthin, and hydroxycinnamic acid. Hesperetin, for example, has been demonstrated to exert cancer chemopreventative benefits in lung and colon tissues, largely through its impact on modulation of phase I and phase II enzyme activity.[48–50] In studies of a rat model of chemically induced colon cancer, expression of a range of detoxification enzymes, including SOD, CAT, GPx, GR, reduced GSH, GST, and NQO1, was significantly higher in the hesperetin group compared to the control groups. The hesperetin-fed group had attenuated levels of lipid peroxidation and, most dramatically, developed less dysplasia and significantly fewer tumors than the nontreated groups.[48,49] Similar increases in detoxification enzymes were observed in a mouse model of lung cancer.[50]

Blood (red) oranges, common in the Mediterranean diet, are unique in that, in addition to the typical citrus bioactives, they contain high concentrations of anthocyanins (e.g., cyanidin 3-glucoside) giving them a bright purple (blood) juice color. Both the juice (BOJ) and the dried juice extract (BOE) of blood oranges have been studied *in vitro* and *in vivo*, including human intervention trials. There is good evidence that the positive effects of BOJ are a result of the full complex of phytonutrients, rather than a benefit of vitamin C alone. When subjects consumed a drink with vitamin C, versus BOJ containing an equal amount of vitamin C, protective effects against DNA damage were only observed in the BOJ group.[51] Peripheral blood mononuclear cells from the BOJ group were more resistant to *ex vivo* H_2O_2-induced DNA strand breaks, indicating that the full complex was more efficacious than at least one of the individual components.[51] In diabetic subjects, an established oxidative stress condition, consumption of 100 mg/day of BOE for two months led to increased levels of serum protein thiols, an indirect measure of GSH activity, and those subjects with the highest levels of lipid peroxidation at baseline, had measurable decreases by the end of two months.[52] BOJ and BOE have been specifically demonstrated to protect against occupational and recreational stressors. Bonina *et al.*[53] reported that 100 mg/day of a BOE for one month protected traffic police officers against air pollution–induced oxidative damage. In only one month, improvements were observed in markers of lipid peroxidation, DNA damage, and GSH status. High-intensity physical activity is another known source of oxidative damage, especially for those individuals at the extremes of the continuum: sedentary individuals unaccustomed to exercise and professional athletes. BOE has been demonstrated to protect against oxidative stress in the latter group. Professional handball players supplemented with 100 mg/day BOE for two months benefited from lower levels of lipid peroxidation, higher GSH activity, and an attenuation of DNA damage.[54] *In vitro* studies suggest that BOE has potent anti-inflammatory activities at the levels of gene and protein expression.[55–57] For example, Frasca *et al.*[57] studied the anti-inflammatory effects of BOE in human articular chondrocytes, a model of osteoarthritis, and found that BOE downregulated the expression of a number of inflammatory genes at the transcriptional level, including iNOS, COX-2, and intercellular adhesion molecule 1. They observed corresponding decreases in the abundance of the proinflammartory cytokines prostaglandin E_2 and IL-8 as well as nitrite, which is known to contribute to the loss of joint cartilage.[57]

The mechanisms of action of the blood orange are not well elucidated, but it is likely that it acts through the modulation of gene expression associated with intrinsic antioxidants, vitagenes, and/or anti-inflammatory pathways. The benefits elicited by BOJ and BOE in humans, despite the relatively low doses of the supplements, indicates that the citrus bioactives are more likely to be acting indirectly via a hormetic action, rather than directly as antioxidants. The broad protective effects of BOE at relatively low doses against environmental and lifestyle-associated stressors, make this ingredient an attractive candidate for human antiaging research.

Summary

A conclusion, based on the positive effects of these phytochemicals on the expression of key genes, including vitagenes associated with cellular detoxification, antioxidant protection, and inflammatory balance in various tissues *in vivo*, was made that it is possible to design a blend of materials to specifically target changes in aging-associated pathways. Individual nutrients had greater or lesser effects on specific Nrf2- and inflammation-related genes in various tissues and experimental models. Therefore, creating a blend of ingredients designed to modulate the greatest diversity of Nrf2-and inflammation-related genes in the greatest number of tissues would likely have the most dramatic protective effects against oxidative damage, toxicants, and inflammation, providing the most robust antiaging benefits.

Acknowledgments

The authors would like to thank Jamie L. Barger, Tomas A. Prolla, and Richard Weindruch at LifeGen Technologies for sharing their expertise in the areas of gene expression and aging. We would also like to thank members of the Nu Skin Enterprises Center for Anti-Aging Research for helpful scientific discussion: Joseph Chang, Mark Bartlett, Doug Burke, Stephen Poole, Scott Ferguson, Glenn Cheney, Josh Zhu, Lih-Wen Ding, and Bill Landreth.

Conflicts of interest

Both authors are employed by Nu Skin Enterprises, a dietary supplement and personal care company.

References

1. Han, X., T. Shen & H. Lou. 2007. Dietary polyphenols and their biological significance. *Int. J. Mol. Sci.* **8:** 950–988.
2. Lotito, S. & B. Frei. 2006. Consumption of flavonoid-rich foods and increased plasma antioxidant capacity in humans: cause, consequence, or epiphenomenon? *Free Radical Bio. Med.* **41:** 1727–1746.
3. Shih, P. & G. Yen. 2007. Differential expressions of antioxidant status in aging rats: the role of transcriptional factor Nrf2 and MAPK signaling pathway. *Biogerontology* **8:** 71–80.
4. Zimniak, P. 2008. Detoxification reactions: relevance to aging. *Ageing Res. Rev.* **7:** 281–300.
5. Calabrese, V. *et al.* 2012. Cellular stress responses, hormetic phytochemicals and vitagenes in aging and longevity. *Biochim. Biophys. Acta* **1822:** 753–783.
6. Calabrese, V. *et al.* 2011. Hormesis, cellular stress response and vitagenes as critical determinants in aging and longevity. *Ageing Res. Rev.* **32:** 279–304.
7. Halliwell, B. & J. Gutteridge. 2007. *Free Radicals in Biology and Medicine.* Oxford University Press. Oxford.
8. Suh, J. *et al.* 2004. Decline in transcriptional activity of Nrf2 causes age-related loss of glutathione synthesis, which is reversible with lipoic acid. *Proc. Natl. Acad. Sci. USA* **101:** 3381–3386.
9. Toroser, D. & R. Sohal. 2007. Age-associated perturbations in glutathione synthesis in mouse liver. *Biochem. J.* **405:** 583–589.
10. Luchak, J., L. Prabhudesai, R. Sohal, *et al.* 2007. Modulating longevity in Drosophila by over- and underexpression of glutamate-cysteine ligase. *Ann. N.Y. Acad. Sci.* **1119:** 260–273.
11. Liu, H., H. Wang, S. Shenvi, *et al.* 2004. Glutathione metabolism during aging and in Alzheimer disease. *Ann. N.Y. Acad. Sci.* **1019:** 346–349.
12. Lu, S. 2002. Regulation of glutathione synthesis. *Mol. Aspects Med.* **30:** 42–59.
13. McElwee, J. *et al.* 2007. Evolutionary conservation of regulated longevity assurance mechanisms. *Genome Biol.* **8:** R132.1–R132.16.
14. Hyun, D., J. Hernandez, M. Mattson & R. de Cabo. 2006. The plasma membrane redox system in aging. *Ageing Res. Rev.* **5:** 209–220.

15. Brooks, J., V. Paton & G. Vidanes. 2001. Potent induction of phase 2 enzymes in human prostate cells by sulforaphane. *Cancer Epidem. Biomar.* **10:** 949–954.
16. Leiser, S. & R. Miller. 2010. Nrf2 Signaling, a mechanism for cellular stress resistance in long-lived mice. *Mol. Cell Biol.* **30:** 871–884.
17. Collins, A. *et al.* 2009. Age-accelerated atherosclerosis correlates with failure to upregulate antioxidant genes. *Circ. Res.* **104:** e42–e54.
18. Tanito, M. *et al.* 2005. Sulforaphane induces thioredoxin through the antioxidant-responsive element and attenuates retinal light damage in mice. *Invest. Ophth. Vis. Sci.* **46:** 979–987.
19. Donato, A., A. Black, K. Jablonski, *et al.* 2008. Aging is associated with greater nuclear NFkB, reduced IkBalpha, and increased expression of proinflammatory cytokines in vascular endothelial cells of healthy humans. *Aging Cell* **7:** 805–812.
20. Kang, M. *et al.* 2005. The effect of age and calorie restriction on HIF-1-responsive genes in aged liver. *Biogerontology* **6:** 27–37.
21. Prisby, R. *et al.* 2007. Aging reduces skeletal blood flow, endothelium-dependent vasodilation, and NO bioavailability in rats. *J. Bone Mineral Res.* **22:** 1280–1288.
22. Gendron, M. & E. Thorin. 2007. A change in the redox environment and thromboxane A2 production precede endothelial dysfunction in mice. *Am. J. Physiol.* **293:** H2508–H2515.
23. Naik, A. *et al.* 2009. Reduced COX-2 expression in aged mice is associated with impaired fracture healing. *J. Bone Mineral Res.* **24:** 251–264.
24. Shapiro, T. *et al.* 2006. Safety, tolerance, and metabolism of broccoli sprout glucosinolates and isothiocyanates: a clinical phase I study. *Nutr. Cancer* **55:** 53–62.
25. Fahey, J. & P. Talalay. 1999. Antioxidant functions of sulforaphane: a potent inducer of phase II detoxication enzymes. *Food Chem. Toxicol.* **37:** 973–979.
26. Thimmulappa, R. *et al.* 2002. Identification of Nrf2-regulated genes induced by the chemopreventive agent sulforaphane by oligonucleotide microarray. *Cancer Res.* **62:** 5196–5203.
27. Hu, R. *et al.* 2006. Gene expression profiles induced by cancer chemopreventive isothiocyanate sulforaphane in the liver of C57BL/6J mice and C57BL/6J/Nrf2 (–/–) mice. *Cancer Lett.* **243:** 170–192.
28. Xue, M. *et al.* 2008. Activation of NF-E2-related factor-2 reverses biochemical dysfunction of endothelial cells induced by hyperglycemia linked to vascular disease. *Diabetes* **57:** 2809–2817.
29. Angeloni, C. *et al.* 2009. Modulation of phase II enzymes by sulforaphane: implications for its cardioprotective potential. *J. Agric. Food Chem.* **57:** 5615–5622.
30. Mizuno, K. *et al.* 2011. Glutathione biosynthesis via activation of the nuclear factor E2-related factor 2 (Nrf2) – antioxidant-response element (ARE) pathway is essential for neuroprotective effects of sulforaphane and 6-(methylsulfinyl) hexyl isothiocyanate. *J. Pharmacol. Sci.* **115:** 320–328.
31. Hong, Y., W. Yan, S. Chen, *et al.* 2010. The role of Nrf2 signaling in the regulation of antioxidants and detoxifying enzymes after traumatic brain injury in rats and mice. *Acta Pharm. Sinic.* **31:** 1421–1430.
32. Malaguti, M. *et al.* 2009. Sulforaphane treatment protects skeletal muscle against damage induced by exhaustive exercise in rats. *J. Appl. Physiol.* **107:** 1028–1036.
33. Mukherjee, S., H. Gangopadhyay & D. Das. 2007. Broccoli: a unique vegetable that protects mammalian hearts through the redox cycling of the thioredoxin superfamily. *J. Agric. Food Chem.* **56:** 609–617.
34. Gasper, A. *et al.* 2005. Glutathione S-transferase M1 polymorphism and metabolism of sulforaphane from standard and high-glucosinolate broccoli. *Am. J. Clin. Nutr.* **82:** 1283–1291.
35. Gasper, A. *et al.* 2007. Consuming broccoli does not induce genes associated with xenobiotic metabolism and cell cycle control in human gastric mucosa. *J. Nutr.* **137:** 1718–1724.
36. Riedl, M., A. Saxon & D. Diaz-Sanchez. 2009. Oral sulforaphane increases phase II antioxidant enzymes in the human upper airway. *Clin. Immunol.* **130:** 244–251.
37. Petri, N. *et al.* 2003. Absorption/metabolism of sulforaphane and quercetin, and regulation of phase II enzymes, in human jejunum *in vivo*. *Drug Metab. Dispos.* **31:** 805–813.
38. Gill, C. *et al.* 2004. The effect of cruciferous and leguminous sprouts on genotoxicity, *in vitro* and *in vivo*. *Cancer Epidem. Biomar.* **13:** 1199–1205.
39. Murashima, M., S. Watanabe, X. Zhuo, *et al.* 2004. Phase 1 study of multiple biomarkers for metabolism and oxidative stress after one-week intake of broccoli sprouts. *BioFactors* **22:** 271–275.
40. Terra, X. *et al.* 2009. Grape-seed procyanidins prevent low-grade inflammation by modulating cytokine expression in rats fed a high-fat diet. *J. Nutr. Biochem.* **20:** 210–218.
41. Terra, X. *et al.* 2011. Modulatory effect of grape-seed procyanidins on local and systemic inflammation in diet-induced obesity rats. *J. Nutr. Biochem.* **22:** 380–387.
42. Wang, Y. *et al.* 2009. Consumption of grape seed extract prevents amyloid-beta deposition and attenuates inflammation in brain of an Alzheimer's disease mouse. *Neurotox. Res.* **15:** 3–14.
43. Sano, A. *et al.* 2007. Beneficial effects of grape seed extract on malondialdehyde-modified LDL. *J. Nutr. Sci. Vitaminol.* **53:** 174–182.
44. Sivaprakasapillai, B., I. Edirisinghe, J. Randolph, *et al.* 2009. Effect of grape seed extract on blood pressure in subjects with the metabolic syndrome. *Metabolism* **58:** 1743–1746.
45. El-Alfy, A., A. Ahmed & A. Fatani. 2005. Protective effect of red grape seeds proanthocyanidins against induction of diabetes by alloxan in rats. *Pharmacol. Res.* **52:** 264–270.
46. Kar, P., D. Laight, H. Rooprai, *et al.* 2009. Effects of grape seed extract in type 2 diabetic subjects at high cardiovascular risk: a double blind randomized placebo controlled trial examining metabolic markers, vascular tone, inflammation, oxidative stress and insulin sensitivity. *Diabetic Med.* **26:** 526–531.
47. Vigna, G. *et al.* 2003. Effect of a standardized grape seed extract on low-density lipoprotein susceptibility to oxidation in heavy smokers. *Metabolism* **52:** 1250–1257.

48. Aranganathan, S., J. Panneer Selvam, N. Sangeetha & N. Nalini. 2009. Modulatory efficacy of hesperetin (citrus flavanone) on xenobiotic-metabolizing enzymes during 1,2-dimethylhydrazine-induced colon carcinogenesis. *Chem.-Biol. Interact.* **180:** 254–261.
49. Aranganathan, S. & N. Nalini. 2009. Efficacy of the potential chemopreventive agent, hesperetin (citrus flavanone), on 1,2-dimethylhydrazine induced colon carcinogenesis. *Food Chem. Toxicol.* **47:** 2594–2600.
50. Kamaraj, S., G. Ramakrishnan, P. Anandakumar, *et al.* 2009. Antioxidant and anticancer efficacy of hesperidin in benzo(a)pyrene induced lung carcinogenesis in mice. *Invest. New Drug.* **27:** 214–222.
51. Guanieri, S., P. Riso & M. Porrini. 2011. Orange juice vs. vitamin C: effect on hydrogen peroxide-induced DNA damage in mononuclear blood cells. *Brit. J. Nutr.* **97:** 639–643.
52. Bonina, F. *et al.* 2002. Evaluation of oxidative stress in diabetic patients after supplementation with a standardised red orange extract. *Diabetes Nutr. Metab.* **15:** 14–19.
53. Bonina, F. *et al.* 2008. Protective effects of a standardised red orange extract on air pollution-induced oxidative damage in traffic police officers. *Nat. Prod. Res.* **22:** 1544–1551.
54. Bonina, F. *et al.* 2005. Oxidative stress in handball players: effect of supplementation with a red orange extract. *Nutr. Res.* **25:** 917–924.
55. Cardile, V., G. Frasca, L. Rizza, *et al.* 2010. Antiinflammatory effects of a red orange extract in human keratinocytes treated with interferon-gamma and histamine. *Phytother. Res.* **24:** 414–418.
56. Cimino, F., M. Cristani, A. Saija, *et al.* 2007. Protective effects of a red orange extract on UVB-induced damage in human keratinocytes. *BioFactors* **30:** 129–138.
57. Frasca, G. *et al.* 2010. Involvement of inducible nitric oxide synthase and cyclooxygenase-2 in the anti-inflammatory effects of a red orange extract in human chondrocytes. *Nat. Prod. Res.* **24:** 1469–1480.

Table 2. *MeCP2* **mutations are not synonymous with RTT**[a]

MeCP2 mutation	RTT	Clinical presentation	Biological significance
+	+	Typical RTT Preserved speech variant (PSV)	*MeCP2* mutation is the major cause of typical RTT
−	+	Early seizure variant (ESV) Congenital variant (CV) *MeCP2*-negative RTT-like phenotypes	Mutations in other genes are mainly associated with atypical RTT
+	−	Asymptomatic female carriers[35] Neonatal hypoxic encephalopathy in males[36] *MeCP2* duplication[124] Autism[37] Angelman syndrome–like presentation[38] Nonspecific intellectual disability[125] X-linked intellectual disability[126]	Emerging concept of *MeCP2*-pathies (MeCP2-related syndromes)

[a]More in-depth information regarding MeCP2 mutation in non-RTT syndromes can be found in Ref. 142.

females are the true model for RTT, but most mouse studies have used hemizygous null males, owing to their earlier symptom onset and homogeneous population of MeCP2-negative cells. Despite great advantages over patient samples, the study of MeCP2 in mouse models still provides several challenges.[1] However, in our opinion, the most important information derived by the animal models is the demonstration, by Guy *et al.* in 2007,[41] of disease reversibility (Table 1). This crucial finding, although so far confined to RTT animal models, when coupled with the lack of neuronal necrosis in the disease[42] strongly emphasizes the reversibility of the pathogenic mechanisms underlying the disease.

Link between oxidative stress and RTT

Reactive oxygen species (ROS) are the physiologic by-products of several essential biological processes, although they are known to be potentially damaging to cells. Eukaryotic organisms have evolved a comprehensive range of proteins to detoxify ROS and repair against oxidative damage to DNA, lipids, or proteins. Antioxidants include enzymatic scavengers, such as superoxide dismutase (SOD) and catalase, glutathione peroxidase, and peroxyredoxins, as well as nonenzymatic factors, including glutathione, flavonoids, and vitamins.[43,44] Cumulating evidence indicates that OS is not only deleterious but, in certain circumstances, also serves critical developmental and/or physiological processes.[45] In technical terms, OS occurs when the antioxidant response is insufficient to balance the production of ROS. This state can ultimately cause cell death by apoptosis or necrosis via an array of signaling pathways. To date, many studies have demonstrated that normal functioning of antioxidant defense systems is essential for cell survival,[46] as mouse knockouts of the most critical mitochondrial antioxidant genes (including glutathione peroxidase 4, thioredoxin 2, and SOD2)[47–49] often lead to early lethality. Recently, the role of ROS in neurons has been a particular focus of research due to the consistent presence of various OS markers in neurodegenerative disease, as well as several pathogenic mutations in proteins prominently featured in antioxidant pathways.[46]

The apparent discrepancy between the lack of neuronal degeneration in the brain of RTT patients and the prominence of the functional abnormalities should be emphasized. This point has been well expressed by professor Adrian Bird in a scientific interview.[42] Moreover, MeCP2 has been implicated in cell commitment and maintenance in neurons by triggering senescence.[50] On the other hand, there is an emerging body of evidence that underscores the importance of the damage in the astrocytes and microglia. Neurotoxic activity, consequent to an excessive release of glutamate, has been reported in MeCP2-deficient microglia. Emerging evidence strongly suggests that the expression of MeCP2 in glia is more ubiquitous than originally thought, and that MeCP2 deficiency in glia may have a profound impact on brain function, building up the concept that glial cells, like neurons, are integral components of the neuropathology of RTT.[51–53]

The brain is the most vulnerable to ROS damage compared to other organs due to high concentrations of unsaturated lipids, a high metabolic rate, high levels of cellular iron, and relatively low levels of potentially protective enzymes and free radical scavengers.

It has been known for a long time[54] that redox-active iron is one of the most active sources of OS. Our prior research demonstrates that iron is released in a free redox-active form (nonprotein bound iron, NPBI) when erythrocytes are challenged by OS, either induced by hemolytic drugs,[55] or generated in a more subtle way, such as the shift from low oxygen pressure in utero to high oxygen pressure that occurs at birth.[55] Subsequently, the released iron can diffuse out of the erythrocytes and the diffusion is higher with hypoxic erythrocytes.[56] Iron is present in some areas of the brain in relatively high concentrations[57] and, to be redox-cycling active, must to be released in a "free" form from its macromolecular complexes.[58] The iron chelator desferrioxamine (DFO) has been shown to be able to prevent neuronal injury in animal models of cerebral ischemia.[59]

Currently, there is unanimous consensus about the reliability of isoprostanes (IsoPs), a class of oxidation products derived from nonenzymatic peroxidation of several polyunsaturated fatty acids (PUFAs) precursors, as reliable markers of OS *in vivo.* IsoPs are chemically stable compounds that, with proper techniques, are measurable in an accurate, reproducible, and reliable way down to the order of picograms.

In 1990, Morrow *et al.* first reported that IsoPs are generated *in vivo* during an OS from arachidonic acid (AA, C20:4 ω-6), via a free radical–catalyzed mechanism.[60] OS has been implicated in a wide variety of human disorders such as diabetes, pulmonary, cardiovascular or neurodegenerative diseases. Furthermore, IsoPs are commonly used in clinical trials as a reliable OS biomarkers in many diseases and pathologies.[61] However, it is increasingly clear that besides serving as reliable OS markers, IsoPs possess their own biological activities.[62,63]

In 1998, a novel class of IsoPs named neuroprostanes (NeuroPs) was discovered.[64,65] The name of neuroprostane was assigned because these NeuroPs are generated from docosahexaenoic acid (DHA, C22:6 ω-3), which is among the most abundant PUFAs in both brain and retina and is an essential requirement for their development.[66] Levels of F_4-NeuroPs are 2.1-fold higher in temporal lobe of Alzheimer disease patients than in healthy controls[67] and are fourfold higher than F_2-IsoPs levels.[68]

At the same time, novel IsoPs derived from eicosapentaenoic acid (EPA, C20:5 ω-3) were discovered and named F_3-IsoPs.[69] Additionally, it was found that F_3-IsoPs could be generated from F_4-NeuroPs, following a β-oxidation process.[70]

The last discovered family of IsoPs was observed through adrenic acid (AdA, C22:4 ω-6) peroxidation.[71] Similarly to DHA, AdA is concentrated in the brain and especially in myelin within white matter. Whereas DHA leads to the formation of F_4-NeuroPs, AdA leads to F_2-dihomo-IsoPs, whose levels have been reported to be significantly increased in brain white matter samples from Alzheimer disease patients.[71] Consequently, F_4-NeuroPs and F_2-dihomo-IsoPs can be considered powerful tools for gauging the oxidative damage of the central nervous system *in vivo.*

In parallel with the discoveries in molecular biology and the clinical features of RTT, a possible causative role for OS in the disease pathogenesis has been repeatedly suggested since at least 1987, that is, 21 years after the first clinical RTT description. Interestingly, Rett was among the authors of this study that indicated reduced ascorbic acid and glutathione in a postmortem brain study from a single patient.[72] However, this observation was reported 11 years before reduced antioxidant defense (low vitamin E concentrations) could be demonstrated in the serum of RTT patients, 14 years before an oxidative imbalance could be reported,[73] and over 22 years before lipid and protein oxidative damage could be demonstrated in patients with typical RTT[74] (Table 1). The current knowledge of the link between OS and RTT is reviewed in Table 3, in which a complex constellation of deficits in the antioxidant defense system and excesses in ROS generation appear to be evident.

Nevertheless, the first direct demonstration of a link between MeCP2 and OS could be traced back to 2004, when Valinluck *et al,* reported experimental evidence that oxidative damage to methyl-CpG sequences is able to reduce the binding affinity of the MBD of MeCP2 by at least an order of magnitude,[75] thus suggesting that oxidative damage to DNA could result in heritable, epigenetic changes in

Table 3. The link between OS and RTT: summary of the current knowledge

Antioxidant defenses	References
Antioxidant enzyme activities	
↓ Erythrocyte SOD	73
↔ Erythrocyte glutathione peroxidase	
↔ Erythrocyte glutathione reductase	
↔ Erythrocyte catalase	
Glutathione	
↓ Brain GSH (postmortem)	72
Vitamins	
↓ Brain vitamin C (postmortem)	72
↓ Serum vitamin E	109
Gene expression (postmortem brain)	
↑ Catalase	127
↑ Microsomal glutathione S-transferase	
↑ Glutathione S-transferase π	

Oxidative stress markers	References
Pro-oxidant factors	
↑ Plasma NPBI	74
↑ Erythrocyte NPBI	94
↑ Serum global OS (nmol/l H_2O_2)	128
Lipid peroxidation markers	
↑ Plasma MDA	73
↑ Plasma F_2-IsoPs	74
	94
↑ Plasma F_4-NeuroPs	93
↑ Plasma F_2-dihomo-IsoPs	91
↑ Erythrocyte F_2-IsoPs	90
Protein oxidation markers	
↑ Plasma protein carbonyls	74
↑ Plasma 4-HNE PAs	129
↑ Erythrocyte 4-HNE PAs	90
Mitochondrial alteration/dysfunction	
↑ CSF lactate	130, 131
↑ Blood lactate	130, 132
↑ CSF pyruvate	131
Morphological alterations (human tissues)	133
Morphological alterations (muscle biopsies)	134–136
Enzyme activities	
↓ Cyt C oxidase, succinate cyt C reductase activities (muscle biopsies)	137
↓ NADH cyt C reductase, succinate cyt C reductase activities (muscle biopsies)	138
↔ Pyruvate dehydrogenase, citrate synthetase activities (skin fibroblasts)	132
↔ Cyt C oxidase activity (muscle biopsies)	138
Gene expression	
↓ Malate dehydrogenase and succinate dehydrogenase	127
↓ Cyt C oxidase subunit 1 (brain in MeCP2 null mice)	80
↓ Cyt C oxidase subunit 1 (frontal cortex of post-mortem brain biopsy)	139

↑ increased; ↔ unchanged; ↓ decreased.

chromatin organization, and indicating that MeCP2 can be both the starting point and one of the end targets of a kind of chain reaction process triggered by free radical species.

In addition, *MeCP2* has been reported to play a role in the regulation of molecular targets involved in the adaptive response to OS, such as brain-derived neurotrophic factor (BDNF),[76] cAMP responsive element-binding protein (CREB),[77] and proline dehydrogenase (*Prodh*),[78] among many others likely not yet recognized. This notion is also in line with previous reports on mitochondrial abnormalities in RTT patients[79] and experimental animal models.[80]

Oxidative stress and Rett syndrome: personal contribution

Since 2009, our multidisciplinary team—including physicians, physiologists, pathophysiologists, molecular geneticists, molecular biologists, and chemists—has pursued the OS hypothesis in the search for the mechanisms behind RTT in an attempt to fill the apparent gap between the occurrence of *MeCP2* gene mutation and disease expression as a function of the time (clinical stages) and of phenotype severity (different clinical forms, including typical and atypical RTT). We have tried to investigate OS by measuring biomarkers that could be intimately involved in a common mechanism, starting from the action of prooxidant factors, ultimately leading to the oxidation of several biological macromolecules essential for proper cellular maintenance and function.

From a clinical point of view, we have firstly described the occurrence of a chronic mild hypoxia with impaired lung gas exchange, in association with clear evidence of OS damage in typical RTT patients.[74] The spectrum of reported breathing irregularities in RTT include breath holding, spontaneous Valsalva maneuvers, apnea, apneusis, hyperventilation, and rapid shallow breathing.[81–87] These respiratory disorders have been reported during wakefulness as well as during the nighttime[86] and are commonly attributed to autonomic dysfunction and/or brain-stem immaturity.[87]

However, although a primary brainstem dysfunction might be able to explain some of the breathing dysfunctions (abnormal breathing rate and rhythm; central apneas) from patients and knockout-engineered mice, it cannot fully explain some relevant breathing features, including obstructive apneas,[88,89] as well as the impaired pulmonary gas exchanges detectable in RTT,[74] which recently we have found to be related to abnormal blood red cell shapes.[90]

Therefore, chronic hypoxia could be a possible source for hemoglobin autoxidation and the subsequent increase in the NPBI catalyzing the redox reactions. However, to date, many pieces of the puzzle are still missing in the unexplored "black box" between the *MeCP2* gene mutation and the associated OS derangement, as well as a clear experimental proof for a cause–effect relationship.

The results of our investigations on OS in RTT are summarized in Tables 4 and 5. Our findings indicate that typical RTT is characterized by markedly increased levels of IsoPs deriving from the nonenzymatic oxidation of AA, DHA, and AdA—all of the stages. In particular, we found extremely high (two orders of magnitude) plasma levels of F_2-dihomo-IsoPs, AdA oxidation products, in RTT girls in stage I of the disease, a stage characterized by a dramatic neurologic regression that is one of the major hallmarks of this neurologic disease.[91] AdA, whatever the actual origin (brain white matter, adrenal gland, or kidney), is the PUFA that goes through the greatest degree of oxidation in the earliest stage of the typical form of the disease. Our finding suggests that brain oxidative damage already occurs during the first two years of the natural evolution of the disease, and that this damage also involves the peroxidation of AdA, a critical component of myelin[71] in the primate brain. Oligodendrocytes synthesize large amounts of plasma membrane and wrap it around axons in the CNS. One main function of myelin is to insulate the axon and to cluster sodium channels into the nodes of Ranvier, and this, in turn, enables the action potential to jump from one node to the other.[92] Our results in early RTT (stage I of the typical disease) indicate that an early insult to AdA corresponded with greatly increased F_2-dihomo-IsoPs plasma levels in the patients, coinciding with the clinical onset of neuroregression, and is the prominent lipid peroxidation end product detectable at this stage.[91]

The natural course of the disease is also accompanied by a moderate to marked increase in NPBI, a prooxidant factor causing release of hydroxyl radicals as the result of the Fenton reaction, while

Table 4. OS markers in typical and atypical RTT

		Plasma isoprostanes			NPBI			
		F_2-IsoPs	F_4-NeuroPs	F_2-dihomo IsoPs	Plasma	Intraerythrocyte	Plasma 4-HNE PAs	Main clinical features/ phenotype
Typical RTT								
Clinical stages	I	↑↑↑	↑↑↑	↑↑↑	↑↑	↑↑	↔	Neurological regression
	II	↑↑↑	↑↑↑	↑↑↑	↑↑	↑↑↑	↑↑	Autistic behavior
	III	↑↑	↑↑↑	↑↑↑	↑↑	↑↑↑	↑↑↑	"Pseudo-stationary"
	IV	↑↑	↑↑↑	↑↑↑	↑↑	↑↑↑	↑↑↑	"Wheel-chair stage"
Atypical RTT								
Early seizure variant (ESV)		↑↑↑	↑	N.D.	↑↑	↑↑↑	↑↑↑	Early onset of seizures
Preserved speech variant (PSV)		↔	↑↑	N.D.	↑	↔	↑↑↑	Late regression; recovery of language after regression; autistic behavior; IQ up to 50; normal head circumference
Congenital variant (CV)		↑	N.D.	N.D.	↑	↑↑	↔	Early regression; severe psychomotor delay; severe early microcephaly; specific movement abnormalities (tongue; jerky movements of limbs)

↔ unchanged; ↑ mild increase (1.0 < Z-Score ≤1.5 or 1.0 < MoMs ≤1.5); ↑↑ moderate increase (1.6 ≤ Z-Score ≤3 or 1.6 ≤ MoMs <3); ↑↑↑ severe increase (>3 Z-Score or >3 MoMs); N.D., not determined.

evidence of oxidative protein damage is apparent from clinical stage II.

In addition, OS marker levels appear to be related to phenotype severity. In particular, OS is consistently enhanced in the patients with the ESV, while it appears to be generally milder in the PSV and congenital variant patients (Table 4).

Moreover, *MeCP2* gene mutations located in critical regions that carry higher phenotype severity usually show more shifted OS imbalance (Table 5). We have previously shown that the degree of MeCP2 protein dysfunction is directly proportional to the OS-mediated neuronal damage, explaining ~90% of the expressed phenotype variability.[93]

Table 5. OS markers as a function of *MeCP2* gene mutations type

	Plasma isoprostanes		NPBI			
MeCP2 gene mutation	F_2-IsoPs	F_4-NeuroPs	Plasma	Intraerythrocyte	Phenotype severity	References
R106W	↔	↑↑↑	↑↑	↑↑	++	140
R133C	↑	↑	↑↑	↑↑	+	30–32
T158M	↑↑↑	↑↑↑	↑↑	↑↑↑	+/++	23 141 140 31
R168X	↑↑	↑↑↑	↑↑	↑↑↑	+++	23, 31–33
R255X	↑↑	↑↑↑	↑↑↑	↑↑↑	++/+++	30 140
R270X	↑↑	↑↑↑	↑↑↑	↑↑↑	+++	30
R294X	↔	↑	↑↑	↑↑	+	23 30 31
R306C	↔	↑↑	↑↑	↑↑	+ / ++	23 140
C-term deletions	↑	↑↑	↑↑	↑↑↑	+	23
Large deletions	↑	↑↑↑	↑↑	↑↑↑	+++	23

↔ unchanged; ↑ mild increase ($1.0 < Z$-Score ≤ 1.5 or $1.0 <$ MoMs ≤ 1.5); ↑↑ moderate increase ($1.6 \leq Z$-Score ≤ 3 or $1.6 \leq$ MoMs <3); ↑↑↑ severe increase (>3 Z-Score or >3 MoMs); +mild; ++ moderate; +++ severe.

These findings appear to be compatible with a relevant role for an unbalanced redox status in the pathogenesis of human RTT. A critical confirmation for our oxidative hypothesis might derive from the replication of our results in the available animal models of the disease, focusing on the possible OS changes occurring before the onset of overt neurological symptoms, as these may play a causative role (as opposed to those observed after symptoms onset, which may reflect secondary structural changes, or compensatory effects).

Oxidative stress and potential therapeutic targets

The first foreseeable therapeutic target in RTT, of course, remains the *MeCP2* gene itself. A key milestone in our understanding of this disease is the discovery that RTT can be rescued if the *MeCP2* gene is reexpressed in the brain of the RTT mouse model.[41] Therefore, several strategies for obtaining the *MeCP2* reexpression in the patients were subsequently considered, including activating silent *MeCP2* on inactive X chromosome, use of read-through compounds for nonsense mutations or pharmacological chaperones, and gene therapy. However, several other strategies could be imagined, including manipulation of suppressor genes, regulation of the immune system, or modification of various downstream targets, such as growth factors (BDNF, IGF1), catecholamine signaling (NE, 5HT, DA, MAO inhibitors, reuptake inhibitors, receptor agonists), chromatin remodeling, GABA neurotransmission (GABA agonists), MGlur5 signaling (receptor agonists, AKT/MTOR pathway), the IRAK1/NFkB axis, mitochondria, and bone.

In this context, OS and lipid peroxidation in RTT would serve as major downstream targets, with several available potential candidate molecules with antioxidant or OS regulating activity. ω-3 PUFAs were our first choice given their multiple antioxidant actions and safety of use in the clinical setting. In our prior preliminary studies, exogenous administration of ω-3 PUFAs has been shown to moderately reduce clinical severity[94] and significantly reduce the levels of several OS biomarkers in the blood of the RTT patients.[93,94]

In the near future, further investigation is needed in order to answer several critical open questions regarding the most proper dosage, treatment duration, chemical formulation, organ/cellular fate, a possible active role for their secondary metabolites, and first and foremost, the potential relevance of the time at the onset of the supplementation.

Thus, it is conceivable, as we have shown in early RTT,[91] that early white matter oxidative damage may be—independently from the presence of neurodegene relation—a hallmark of neurological regression potentially applicable to many, if not all, of the major adult neuroregressive syndromes including Alzheimer's disease, multiple sclerosis, and Parkinson's disease. As a consequence, F_2-dihomo-IsoPs may become a common therapeutic target for neuroregressive disorders in the future.

Oxidative stress in RTT: a mechanistic approach

To date, a truly mechanistic explanation for the link between *MeCP2* mutation and OS derangement in RTT girls is still lacking. Human and experimental tests still in progress are reinforcing the vision that the main effect of MeCP2 dysfunction, so far essentially overlooked, translates into a shift of the prooxidant/antioxidant balance toward a hostile prooxidant environment in the harboring patients. Recently, we have demonstrated the beneficial effects (reduction in OS markers levels and improvement of clinical severity scores) of ω-3 PUFAs in the early course of the classical RTT disease.[95] These results strongly reinforce the concept that during the expression of the disease, there is an increased oxidation mainly from AdA, DHA, and EPA. This lipid peroxidation scenario would lead to two predictable biochemical consequences—increased OS and progressive depletion of these PUFAs precursors. Thus, isoprostanes are not only a reliable marker of *in vivo* lipid peroxidation, but may also indicate some of the underlying mechanisms behind the disease pathogenesis. The current knowledge does not allow a full explanation for all the mechanistic details of the likely series of events occurring within the still unexplored black box between MeCP2 dysfunction and disease expression. On the other hand, we strongly believe that the "oxidative approach" to this disease would generate in the future more critical information on the unexpected role of MeCP2 in regulation of OS balance. From the genetic side, an increasing number of genetically determined neurological diseases appear to be mediated by biochemical changes involving increased OS.[96–103] On the OS side, RTT may become a new important model for OS imbalance, as a kind of experiment of nature, readily applicable to more than a dozen of experimental mouse models effectively recapitulating this human disease. Consequently, on one side, the horizon of genetic diseases will be extended to include the biochemical mechanisms mediating phenotype expression, and from the other side, molecular genetics may help to better dissect the chain of biochemical events by applying novel tools, such as the knockout/knockin switches of the function of single genes that specifically target different organs and tissues.

Synopsis

Here, we have shown that the oxidative hypothesis is able to explain several features of this human disease, in particular, its natural history, genotype–phenotype correlation, and clinical heterogeneity. This hypothesis would be compatible with the potential reversibility of the disease previously demonstrated in the RTT animal models. Accumulating evidence indicates that OS biomarkers are related to neurological symptoms severity, mutation type, and clinical presentation. In addition, our findings indicate the importance of blood as a suitable biological fluid for detecting markers of central nervous system oxidative damage in RTT and underline the key role of interaction among organic chemists, OS biochemists, and clinicians in revealing potential new markers of the disease and identifying candidate new targets and interventional strategies aimed at improving the quality of life of these patients, so far affected by an incurable disease.

Further efforts in the near future are needed to answer two major questions in the OS-RTT issue, regarding how and when the OS imbalance is generated. More knowledge must be gathered to better understand the fine molecular mechanisms by which a *MeCP2* mutation can trigger OS imbalance—the black box existing between the *MeCP2* gene mutation and the consequent OS derangement—and the precise timing of onset of the enhanced OS during disease progression.

Conflicts of interest

The authors declare no conflicts of interest.

References

1. Guy, J., H. Cheval, J. Selfridge & A. Bird. 2011. The role of MeCP2 in the brain. *Annu. Rev. Cell Dev. Biol.* **27:** 631–652.
2. Gadalla, K.K., M.E. Bailey & S.R. Cobb. 2011. MeCP2 and Rett syndrome: reversibility and potential avenues for therapy. *Biochem. J.* **439:** 1–14.
3. Percy, A.K. 2011. Rett syndrome: exploring the autism link. *Arch. Neurol.* **68:** 985–989.
4. Shepherd, G.M. & D.M. Katz. 2011. Synaptic microcircuit dysfunction in genetic models of neurodevelopmental disorders: focus on Mecp2 and Met. *Curr. Opin. Neurobiol.* **21:** 827–833.
5. Samaco, R.C. & J.L. Neul. 2011. Complexities of Rett syndrome and MeCP2. *J. Neurosci.* **31:** 7951–7959.
6. Berger-Sweeney, J. 2011. Cognitive deficits in Rett syndrome: what we know and what we need to know to treat them. *Neurobiol. Learn Mem.* **96:** 637–646.
7. Gonzales, M.L. & J.M. LaSalle. 2010. The role of MeCP2 in brain development and neurodevelopmental disorders. *Curr. Psychiatry Rep.* **12:** 127–134.
8. Boggio, E.M., G. Lonetti, T. Pizzorusso & M. Giustetto 2010. Synaptic determinants of Rett syndrome. *Front, Synaptic Neurosci.* **2:** 28.
9. Urdinguio, R.G., J.V. Sanchez-Mut & M. Esteller. 2009. Epigenetic mechanisms in neurological diseases: genes, syndromes, and therapies. *Lancet Neurol.* **8:** 1056–1072.
10. Ricceri, L., B. De Filippis & G. Laviola. 2008. Mouse models of Rett syndrome: from behavioural phenotyping to preclinical evaluation of new therapeutic approaches. *Behav. Pharmacol.* **19:** 501–517.
11. Chahrour, M. & H.Y. Zoghbi. 2007. The story of Rett syndrome: from clinic to neurobiology. *Neuron* **56:** 422–437.
12. Percy, A.K. 2008. Rett syndrome: recent research progress. *J. Child. Neurol.* **23:** 543–549.
13. Rett, A. 1966. Uber ein eigartiges hirnatrophisches Syndrom bei Hyperammoniamie in Kindesalter. *Wien Med. Wochenschr.* **116:** 723–738.
14. Amir, R.E., I.B. Van den Veyver, M. Wan, *et al.* 1999. Rett syndrome is caused by mutation in X–linked MECP2, encoding methyl–CpG–binding protein 2. *Nat. Genet.* **23:** 185–188.
15. Neul, J.L., W.E. Kaufmann, D.G. Glaze, *et al.* 2010. Rett syndrome: revised diagnostic criteria and nomenclature. *Ann. Neurol.* **68:** 944–950.
16. Hagberg, B. 2002. Clinical manifestations and stages of Rett syndrome. *Ment. Retard. Dev. Disabil. Res. Rev.* **8:** 61–65.
17. Chao, H.T., H. Chen, R.C. Samaco, *et al.* 2010. Dysfunction in GABA signalling mediates autism-like stereotypies and Rett syndrome phenotypes. *Nature* **468:** 263–269.
18. Zoghbi, H.Y. 2009. Rett syndrome: what do we know for sure? *Nat. Neurosci.* **12:** 239–240.
19. Miltenberger-Miltenyi, G. & F. Laccone. 2003. Mutations and polymorphisms in the human methyl CpG-binding protein MECP2. *Hum. Mutat.* **22:** 107–115.
20. Christodoulou, J., A. Grimm, T. Maher, *et al.* 2003. RettBASE: the IRSA MECP2 variation database—a new mutation database in evolution. *Hum. Mutat.* **21:** 466–472.
21. D'Esposito, M., N.A. Quaderi, A. Ciccodicola, *et al.* 1996. Isolation, physical mapping and northern analysis of the X-linked human gene encoding methyl CpG-binding protein, MECP2. *Mamm. Gen.* 7: 533–535.
22. Archer, H.L., J. Evans, H. Leonard, *et al.* 2007. Correlation between clinical severity in patients with Rett syndrome with a p.R168X or p.T158M MECP2 mutation, and the direction and degree of skewing of X-chromosome inactivation. *J. Med. Genet.* **44:** 148–152.
23. Neul, J.L., P. Fang, J. Barrish, *et al.* 2008. Specific mutations in methyl-CpG-binding protein 2 confer different severity in Rett syndrome. *Neurology* **70:** 1313–1321.
24. Amir, R.E., I.B. Van den Veyver, R. Schultz, *et al.* 2000. Influence of mutation type and X chromosome inactivation on Rett syndrome phenotypes. *Ann. Neurol.* **47:** 670–679.
25. Cheadle, J.P., H. Gill, N. Fleming, *et al.* 2000. Long-read sequence analysis of the MECP2 gene in Rett syndrome patients: correlation of disease severity with mutation type and location. *Hum. Mol. Genet.* **9:** 1119–1129.
26. Huppke, P., F. Laccone, N. Kramer, *et al.* 2000. Rett syndrome: analysis of MECP2 and clinical characterization of 31 patients. *Hum. Mol. Genet.* **9:** 1369–1375.
27. Hoffbuhr, K., J.M. Devaney, B. LaFleur, *et al.* 2001. MeCP2 mutations in children with and without the phenotype of Rett syndrome. *Neurology* **56:** 1486–1495.
28. Monros, E., J. Armstrong, E. Aibar, *et al.* 2001. Rett syndrome in Spain: mutation analysis and clinical correlations. *Brain Dev.* **23:** S251–S253.
29. Weaving, L.S., S.L. Williamson, B. Bennetts, *et al.* 2003. Effects of MECP2 mutation type, location and X-inactivation in modulating Rett syndrome phenotype. *Am. J. Med. Genet.* **118A:** 103–114.
30. Bebbington, A., A. Anderson, D. Ravine, *et al.* 2008. Investigating genotype-phenotype relationships in Rett syndrome using an international data set. *Neurology* **70:** 868–875.
31. Temudo, T., M. Santos, E. Ramos, *et al.* 2011. Rett syndrome with and without detected MECP2 mutations: An attempt to redefine phenotypes. *Brain Dev.* **33:** 69–76.
32. Downs, J., A. Bebbington, P. Jacoby, *et al.* 2010. Level of purposeful hand function as a marker of clinical severity in Rett syndrome. *Dev. Med. Child Neurol.* **52:** 817–823.
33. Horská, A., L. Farage, G. Bibat, *et al.* 2009. Brain metabolism in Rett syndrome: age, clinical, and genotype correlations. *Ann. Neurol.* **65:** 90–97.
34. Leonard, H., L. Colvin, J. Christodoulou, *et al.* 2003. Patients with the R133C mutation: is their phenotype different from patients with Rett syndrome with other mutations? *J. Med. Genet.* **40:** e52.
35. Wan, M., S.S. Lee, X. Zhang, *et al.* 1999. Rett syndrome and beyond: recurrent spontaneous and familial MECP2 mutations at CpG hotspots. *Am. J. Hum. Genet.* **65:** 1520–1529.

36. Kankirawatana, P., H. Leonard, C. Ellaway, *et al.* 2006. Early progressive encephalopathy in boys and MECP2 mutations. *Neurology* **67:** 164–166.
37. Carney, R.M., C.M. Wolpert, S.A. Ravan, *et al.* 2003. Identification of MeCP2 mutations in a series of females with autistic disorder. *Pediatr. Neurol.* **28:** 205–211.
38. Watson, P., G. Black, S. Ramsden, *et al.* 2001. Angelman syndrome phenotype associated with mutations in MECP2, a gene encoding a methyl CpG binding protein. *J. Med. Genet.* **38:** 224–228.
39. Bahi-Buisson, N., J. Nectoux, H. Rosas-Vargas, *et al.* 2008. Key clinical features to identify girls with CDKL5 mutations. *Brain* **131:** 2647–2661.
40. Ariani, F., J. Hayek, D. Rondinella, *et al.* 2008. FOXG1 is responsible for the congenital variant of Rett syndrome. *Am. J. Hum. Genet.* **83:** 89–93.
41. Guy, J., J. Gan, J. Selfridge, *et al.* 2007. Reversal of neurological defects in a mouse model of Rett syndrome. *Science* **315:** 1143–1147.
42. Gitschier, J. 2009. On the tracks of DNA methylation: an interview to Adrian Bird. *PLoS Genet.* **5:** e1000667.
43. Ott, M., V. Gogvadze, S. Orrenius, *et al.* 2007. Mitochondria, oxidative stress and cell death. *Apoptosis* **12:** 913–922.
44. Calabrese, V., C. Cornelius, C. Mancuso, *et al.* 2008. Cellular stress response: a novel target for chemoprevention and nutritional neuroprotection in aging, neurodegenerative disorders and longevity. *Neurochem. Res.* **33:** 2444–2471.
45. Valko, M., D. Leibfritz, J. Moncol, *et al.* 2007. Free radicals and antioxidants in normal physiological functions and human disease. *Int. J. Biochem. Cell Biol.* **39:** 44–84.
46. Fernandez-Checa, J.C., A. Fernandez, A. Morales, *et al.* 2010. Oxidative stress and altered mitochondrial function in neurodegenerative diseases: lessons from mouse models. *CNS Neurol. Disord. Drug Targets* **9:** 439–454.
47. Yant, L.J., Q. Ran, L. Rao, *et al.* 2003. The selenoprotein GPX4 is essential for mouse development and protects from radiation and oxidative damage insults. *Free Radic. Biol. Med.* **34:** 496–502.
48. Nonn, L., R.R. Williams, R.P. Erickson, *et al.* 2003. The absence of mitochondrial thioredoxin 2 causes massive apoptosis, exencephaly, and early embryonic lethality in homozygous mice. *Mol. Cell Biol.* **23:** 916–922.
49. Lebovitz, R.M., H. Zhang, H. Vogel, *et al.* 1996. Neurodegeneration, myocardial injury, and perinatal death in mitochondrial superoxide dismutase-deficient mice. *Proc. Natl. Acad. Sci. USA* **93:** 9782–9787.
50. Squillaro, T., N. Alessio, M. Cipollaro, *et al.* 2012. Reduced expression of MECP2 affects cell commitment and maintenance in neurons by triggering senescence, new perspective for Rett syndrome. *Mol. Biol. Cell.* **23:** 1435–1445.
51. Maezawa, I. & L.W. Jin. 2010. Rett syndrome microglia damage dendrites and synapses by the elevated release of glutamate. *J. Neurosci.* **30:** 5346–5356.
52. Lioy, D.T., S.K. Garg, C.E. Monaghan, *et al.* 2011. A role for glia in the progression of Rett's syndrome. *Nature* **475:** 497–500.
53. Derecki, N.C., J.C. Cronk, Z. Lu, *et al.* 2012. Wild-type microglia arrest pathology in a mouse model of Rett syndrome. *Nature* **484:** 105–109.
54. Halliwell, B. & J.M.C. Gutteridge. 1985. Oxygen toxicity, oxygen radicals, transition metals and disease. *Biochem. J.* **219:** 1–14.
55. Comporti, M., C. Signorini, G. Buonocore, *et al.* 2002. Iron release, oxidative stress and erythrocyte ageing. *Free Radic. Biol. Med.* **32:** 568–576.
56. Ciccoli, L., V. Rossi, S. Leoncini, *et al.* 2004. Iron release, superoxide production and binding of autologous IgG to band 3 dimers in newborn and adult erythrocytes exposed to hypoxia and hypoxia–reoxygenation. *Biochim. Biophys. Acta* **1672:** 203–213.
57. Papanikolaou, G. & K. Pantopoulos 2005. Iron metabolism and toxicity. *Toxicol. Appl. Pharmacol.* **202:** 199–211.
58. Thomas, C.E. & S.D. Aust. 1985. Rat liver microsomal NADPH-dependent release of iron from ferritin and lipid peroxidation. *Free Radic. Biol. Med.* **1:** 293–300.
59. Shadid, M., G. Buonocore, F. Groenendaal, *et al.* 1998. Effect of deferoxamine and allopurinol on non protein-bound iron concentrations in plasma and cortical brain tissue of newborn lambs following hypoxia ischemia. *Neurosci. Lett.* **248:** 5–8.
60. Morrow, J.D., K.E. Hill, R.F. Burk, *et al.* 1990. A series of prostaglandin F2-like compounds are produced *in vivo* in humans by a non-cyclooxygenase, free radical-catalyzed mechanism. *Proc. Natl. Acad. Sci. USA* **87:** 9383–9387.
61. Kadiiska, M.B., B.C. Gladen, D.D. Baird, *et al.* 2005. Biomarkers of oxidative stress study III. Effects of the nonsteroidal anti-inflammatory agents indomethacin and meclofenamic acid on measurements of oxidative products of lipids in CCl4 poisoning. *Free Radic. Biol. Med.* **38:** 711–718.
62. Jahn, U., T. Durand & J.-M. Galano. 2008. Beyond prostaglandins—chemistry and biology of cyclic oxygenated metabolites formed by free-radical pathways from polyunsaturated fatty acids. *Angew. Chem. Int. Ed.* **47:** 5894–5955.
63. Milne G.L., H. Yin, K.D. Hardy, *et al.* 2011. Isoprostane generation and function. *Chem. Rev.* **111:** 5973–5996.
64. Nourooz-Zadeh, J., E.H.C. Liu, E.E. Änggård, *et al.* 1998. F4-isoprostanes: a novel class of prostanoids formed during peroxidation of docosahexaenoic acid (DHA). *Biochem. Biophys. Res. Commun.* **242:** 338–344.
65. Roberts, L.J., T.J. Montine, W.R. Markesbery, *et al.* 1998. Formation of isoprostane-like compounds (neuroprostanes) *in vivo* from docosahexaenoic acid. *J. Biol. Chem.* **273:** 13605–13612.
66. Sastry, P.S. 1985. Lipids of nervous tissue: composition and metabolism. *Prog. Lipid Res.* **24:** 69–176.
67. Musiek, E.S., J.K. Cha, H. Yin, *et al.* 2004. Quantification of F-ring isoprostane-like compounds (F4-neuroprostanes) derived from docosahexaenoic acid *in vivo* in humans by a stable isotope dilution mass spectrometric assay. *J. Chrom. B: Analyt. Technol. Biomed. Life Sci.* **799:** 95–102.
68. Nourooz-Zadeh, J., E.H. Liu, B. Yhlen, *et al.* 1999. F4-isoprostanes as specific marker of docosahexaenoic acdid peroxidation in Alzheimer's disease. *J. Neurochem.* **72:** 734–740.
69. Nourooz-Zadeh, J., B. Halliwell& E.E. Änggård. 1997. Evidence for the formation of F3-isoprostanes during

peroxidation of eicosapentaenoic acid. *Biochem. Biophys. Res. Commun.* **236:** 467–472.
70. Lawson, J.A., S. Kim, W.S. Powell, *et al.* 2006. Oxidized derivatives of omega-3 fatty acids: identification of IPF3 alpha-VI in human urine. *J. Lipid. Res.* **47:** 2515–2524.
71. VanRollins, M., R.L. Woltjer, H. Yin, *et al.* 2008. F2-dihomo-isoprostanes arise from free radical attack on adrenic acid. *J. Lipid Res.* **49:** 995–1005.
72. Sofić, E., P. Riederer, W. Killian, *et al.* 1987. Reduced concentrations of ascorbic acid and glutathione in a single case of Rett syndrome: a postmortem brain study. *Brain Dev.* **9:** 529–531.
73. Sierra, C., M.A. Vilaseca, N. Brandi, *et al.* 2001. Oxidative stress in Rett syndrome. *Brain Dev. Suppl.* **1:** S236–S239.
74. De Felice, C., L. Ciccoli, S. Leoncini, *et al.* 2009. Systemic oxidative stress in classic Rett syndrome. *Free Radic. Biol. Med.* **47**: 440–448.
75. Valinluck V., H.H. Tsai, D.K. Rogstad, *et al.* 2004. Oxidative damage to methyl-CpG sequences inhibits the binding of the methyl-CpG binding domain (MBD) of methyl-CpG binding protein 2 (MeCP2). *Nucleic Acids Res.* **32:** 4100–4108.
76. Wang, H., G. Yuan, N.R. Prabhakar, *et al.* 2006. Secretion of brain-derived neurotrophic factor from PC12 cells in response to oxidative stress requires autocrine dopamine signaling. *J. Neurochem.* **96:** 694–705.
77. Pugazhenthi, S., K. Phansalkar, G. Audesirk, *et al.* 2006. Differential regulation of c-jun and CREB by acrolein and 4-hydroxynonenal. *Free Radic. Biol. Med.* **40:** 21–34.
78. Urdinguio, R.G., L. Lopez-Serra, P. Lopez-Nieva, *et al.* 2008. Mecp2-null mice provide new neuronal targets for Rett syndrome. *PLoS One* **3:** e3669.
79. Cardaioli, E., M.T. Dotti, J. Hayek, *et al.* 1999. Studies on mitochondrial pathogenesis of Rett syndrome: ultrastructural data from skin and muscle biopsies and mutational analysis at mtDNA nucleotides 10463 and 2835. *J. Submicrosc. Cytol. Pathol.* **31:** 301–304.
80. Kriaucionis, S., A. Paterson, J. Curtis, *et al.* 2006. Gene expression analysis exposes mitochondrial abnormalities in a mouse model of Rett syndrome. *Mol. Cell. Biol.* **26:** 5033–5042.
81. Kerr, A.M. 1992. A review of the respiratory disorder in the Rett syndrome. *Brain Dev. Suppl.* **14:** 43–45.
82. Julu, P.O., A.M. Kerr, S. Hansen, *et al.* 1997. Functional evidence of brain stem immaturity in Rett syndrome. *Eur. Child Adolesc. Psychiatry* **6:** 47–54.
83. Kerr, A.M. & P.O. Julu. 1999. Recent insights into hyperventilation from the study of Rett syndrome. *Arch. Dis. Child.* **80:** 384–387.
84. Julu, P.O., A.M. Kerr, F. Apartopoulos, *et al.* 2001. Characterization of breathing and associated central autonomic dysfunction in the Rett disorder. *Arch. Dis. Child.* **85:** 29–37.
85. Glaze, D.G., J.D. Frost, H.Y. Zoghbi, *et al.* 1987. Rett's syndrome: characterization of respiratory patterns and sleep. *Ann. Neurol.* **21:** 377–382.
86. Weese-Mayer, D.E., S.P. Lieske, C.M. Boothby, *et al.* 2008. Autonomic dysregulation in young girls with Rett syndrome during nighttime in-home recordings. *Pediatr. Pulmonol.* **43:** 1045–1060.
87. Gaultier, C. & J. Gallego. 2008. Neural control of breathing: insights from genetic mouse models. *J. Appl. Physiol.* **104:** 1522–1530.
88. Hagebeuk, E.E., R.P. Bijlmer, J.H. Koelman, *et al.* 2012. Respiratory disturbances in Rett syndrome: don't forget to evaluate upper airway obstruction. *J. Child. Neurol.* Jan 30. [Epub ahead of print].
89. Voituron, N., C. Menuet, M. Dutschmann, *et al.* 2010. Physiological definition of upper airway obstructions in mouse model for Rett syndrome. *Respir. Physiol. Neurobiol.* **173:** 146–156.
90. Ciccoli, L., C. De Felice, E. Paccagnini, *et al.* 2012. Morphological changes and oxidative damage in Rett syndrome erythrocytes. *Biochim. Biophys. Acta* **1820:** 511–520.
91. De Felice, C., C. Signorini, T. Durand, *et al.* 2011. F2-dihomo-isoprostanes as potential early biomarkers of lipid oxidative damage in Rett syndrome. *J. Lipid Res.* **52:** 2287–2297.
92. Aggarwal, S., L. Yurlova & M. Simons. 2011. Central nervous system myelin: structure, synthesis and assembly. *Trends Cell Biol.* **21:** 585–593.
93. Signorini, C., C. De Felice, S. Leoncini, *et al.* 2011. F_4-neuroprostanes mediate neurological severity in Rett syndrome. *Clin. Chim. Acta.* **412:** 1399–1406.
94. Leoncini, S., C. De Felice, C. Signorini, *et al.* 2011. Oxidative stress in Rett syndrome: natural history, genotype, and variants. *Redox Rep.* **16:** 145–153.
95. De Felice, C., C. Signorini, T. Durand, *et al.* 2012. Partial rescue of Rett syndrome by ω-3 polyunsaturated fatty acids (PUFAs) oil. *Genes Nutr.* Mar 8. [Epub ahead of print].
96. Martin, LJ. 2012. Biology of mitochondria in neurodegenerative diseases. *Prog. Mol Biol. Transl. Sci.* **107:** 355–415.
97. Wu, Z.C., J.T. Yu, Y. Li, *et al.* 2012. Clusterin in Alzheimer's disease. *Adv. Clin. Chem.* **56:** 155–173.
98. Corthals, A.P. 2011. Multiple sclerosis is not a disease of the immune system. *Q. Rev. Biol.* **86:** 287–321.
99. Pagano, G. & G. Castello. 2012. Oxidative stress and mitochondrial dysfunction in Down syndrome. *Adv. Exp. Med. Biol.* **724:** 291–299.
100. Perluigi, M. & D.A. Butterfield. 2012. Oxidative stress and down syndrome: a route toward Alzheimer-like dementia. *Curr. Gerontol. Geriatr. Res.* **2012:** 724904.
101. Tiano, L. & J. Busciglio. 2011. Mitochondrial dysfunction and Down's syndrome: is there a role for coenzyme Q(10)? *Biofactors* **37:** 386–392.
102. Campos, C., R. Guzmán, E. López-Fernández, *et al.* 2011. Evaluation of urinary biomarkers of oxidative/nitrosative stress in children with Down syndrome. *Life Sci.* **89:** 655–661.
103. Perluigi, M., F. di Domenico, A. Fiorini, *et al.* 2011. Oxidative stress occurs early in Down syndrome pregnancy: a redox proteomics analysis of amniotic fluid. *Proteomics Clin. Appl.* **5:** 167–178.
104. Hagberg, B., J. Aicardi, K. Dias, *et al.* 1983. A progressive syndrome of autism, dementia, ataxia, and loss of purposeful hand use in girls: Rett's syndrome:report of 35 cases. *Ann. Neurol.* **14:** 471–479.

105. Hanefeld, F. 1985. The clinical pattern of the Rett syndrome. *Brain Dev.* **7:** 320–325.
106. Rolando, S. 1985. Rett syndrome: report of eight cases. *Brain Dev.* **7:** 290–296.
107. Witt Engerström, I. 1992. Age-related occurrence of signs and symptoms in the Rett syndrome. *Brain Dev.* **14**(Suppl): S11–S20.
108. Meehan, R.R., J.D. Lewis, A.P. Bird. 1992. Characterization of MeCP2, a vertebrate DNA binding protein with affinity for methylated DNA. *Nucleic Acids Res.* **20:** 5085–5092.
109. Formichi, P., C. Battisti, M.T. Dotti, *et al.* 1998. Vitamin E serum levels in Rett syndrome. *J. Neurol Sci.* **156:** 227–230.
110. Guideri F., M. Acampa, J. Hayek, *et al.* 1999. Reduced heart rate variability in patients affected with Rett syndrome. A possible explanation for sudden death. *Neuropediatrics* **30:** 146–148.
111. De Bona, C., M. Zappella, J. Hayek, *et al.* 2000. Preserved speech variant is allelic of classic Rett syndrome. *Europ. J. Hum. Genet.* **8:** 325–330.
112. Guy, J., B. Hendrich, M. Holmes, *et al.* 2001. A mouse Mecp2-null mutation causes neurological symptoms that mimic Rett syndrome. *Nat. Genet.* **27:** 322–326.
113. Chen, R.Z., S. Akbarian, M. Tudor, *et al.* 2001. Deficiency of methyl-CpG binding protein-2 in CNS neurons results in a Rett-like phenotype in mice. *Nat. Genet.* **27:** 327–33.
114. Mari, F., S. Azimonti, I. Bertani *et al.* 2005. CDKL5 belongs to the same molecular pathway of MeCP2 and it is responsible for the early-onset seizure variant of Rett syndrome. *Hum. Mol. Genet.* **14:** 1935–1946.
115. Scala, E., F. Ariani, F. Mari, *et al.* 2005. CDKL5/STK9 is mutated in Rett syndrome variant with infantile spasms. *J. Med. Genet.* **42:** 103–107.
116. Chahrour, M., S.Y. Jung, C. Shaw, *et al.* 2008. MeCP2, a key contributor to neurological disease, activates and represses transcription. *Science* **320:** 1224–1229.
117. Gonnelli, S., C. Caffarelli, J. Hayek, *et al.* 2008. Bone ultrasonography at phalanxes in patients with Rett syndrome: a 3-year longitudinal study. *Bone* **42:** 737–742.
118. Neul, J.L., W.E. Kaufmann, D.G. Glaze, *et al.* 2010. Rett syndrome: revised diagnostic criteria and nomenclature. *Ann. Neurol.* **68:** 944–950.
119. De Felice, C., G. Guazzi, M. Rossi, *et al.* 2010. Unrecognized lung disease in classic Rett syndrome: a physiologic and high-resolution CT imaging study. *Chest* **138:** 386–392.
120. Skene P.J., R.S. Illingworth, S. Webb, *et al.* 2010. Neuronal MeCP2 is expressed at near histone-octamer levels and globally alters the chromatin state. *Mol. Cell* **37:** 457–468.
121. Chao, H.T., H. Chen, R.C. Samaco, *et al.* 2010. Dysfunction in GABA signalling mediates autism-like stereotypies and Rett syndrome phenotypes. *Nature* **468:** 263–269.
122. De Felice, C., S. Maffei, C. Signorini, *et al.* 2012. Subclinical myocardial dysfunction in Rett syndrome. Eur. Heart *J. Echocardiogr. Imaging.*
123. Cohen, S., H.W. Gabel, M. Hemberg, *et al.* 2011. Genome-wide activity-dependent MeCP2 phosphorylation regulates nervous system development and function. *Neuron* **72:** 72–85.
124. Ramocki, M.B., S.U. Peters, Y.J. Tavyev, *et al.* 2009. Autism and other neuropsychiatric symptoms are prevalent in individuals with MeCP2 duplication syndrome. *Ann. Neurol.* **66:** 771–782.
125. Donzel-Javouhey, A., C. Thauvin-Robinet, V. Cusin, *et al.* 2006. A new cohort of MECP2 mutation screening in unexplained mental retardation: careful re-evaluation is the best indicator for molecular diagnosis. *Am. J. Med. Genet. A* **140:** 1603–1607.
126. Gomot, M., C. Gendrot, A. Verloes, *et al.* 2003. MECP2 gene mutations in non-syndromic X-linked mental retardation: phenotype-genotype correlation. *Am. J. Med. Genet. A* **123A:** 129–139.
127. Colantuoni, C., O.H. Jeon, K. Hyder, *et al.* 2001. Gene expression profiling in postmortem Rett Syndrome brain: differential gene expression and patient classification. *Neurobiol. Dis.* **8:** 847–865.
128. Carmeli, E., A. Bachar, R. Beiker. Expression of global oxidative stress and matrix metalloproteinases is associated with rett syndrome. 2011. *Neurochemical J.* **5:** 141–145.
129. Pecorelli, A., L. Ciccoli, C. Signorini, *et al.* 2011. Increased levels of 4HNE-protein plasma adducts in Rett syndrome. 2011. *Clin. Biochem.* **44:** 368.
130. Lappalainen, R. & R.S. Riikonen. 1994. Elevated CSF lactate in the Rett syndrome: cause or consequence? *Brain Dev.* **16:** 399–401.
131. Matsuishi, T., F. Urabe, H. Komori, *et al.* 1992. The Rett syndrome and CSF lactic acid patterns. *Brain Dev.* **14:** 68–70.
132. Haas RH., F. Nasirian, X. Hua, *et al.* 1995. Oxidative metabolism in Rett syndrome: 2. Biochemical and molecular studies. *Neuropediatrics* **26:** 95–99.
133. Armostrong, D.D. 1992. The neuropathology of the Rett syndrome. *Brain Dev. Suppl.* **14:** S89–98.
134. Eeg-Olofsson, O., A.G. al-Zuhair, A.S. Teebi, et al. 1989. Rett syndrome: genetic clues based on mitochondrial changes in muscle. *Am. J. Med. Genet.* **32:** 142–144.
135. Ruch, A., T.W. Kurczynski & M.E. Velasco. 1989. Mitochondrial alterations in Rett syndrome . *Pediatr. Neurol.* **5:** 320–323.
136. Dotti, M.T., L. Manneschi, A. Malandrini. *et al.* 1993. Mitochondrial dysfunction in Rett syndrome. An ultrastructural and biochemical study. *Brain Dev.* **15:** 103–106.
137. Coker, S.B. & A.R. Melnyk. 1991. Rett syndrome and mitochondrial enzyme deficiencies. *J. Child. Neurol.* **6:** 164–166.
138. Heilstedt, H.A., M.D. Shahbazian & B Lee. Infantile hypotonia as a presentation of Rett syndrome. 2002. *Am. J. Med. Genet.* **111:** 238–242.
139. Gibson, J.H., B. Slobedman, H. KN, *et al.* 2010. Downstream targets of methyl CpG binding protein 2 and their abnormal expression in the frontal cortex of the human Rett syndrome brain. *BMC Neurosci.* **11:** 53.
140. Glaze, D.G., A.K. Percy, S. Skinner, *et al.* 2010. Epilepsy and the natural history of Rett syndrome. *Neurology* **74:** 909–912.
141. Buoni S., R. Zannolli, C.D. Felice, *et al.* 2008. Drug-resistant epilepsy and epileptic phenotype-EEG association in MECP2 mutated Rett syndrome. *Clin. Neurophysiol.* **119:** 2455–2458.
142. Christodoulou, J. *MECP2-Related Disorders*. Retrieved from: http://www.ncbi.nlm.nih.gov/books/NBK1497/. Last update: April 2, 2009.

Ann. N.Y. Acad. Sci. ISSN 0077-8923

ANNALS OF THE NEW YORK ACADEMY OF SCIENCES
Issue: *Environmental Stressors in Biology and Medicine*

Emerging topics in cutaneous wound repair

Giuseppe Valacchi,[1,2] Iacopo Zanardi,[3] Claudia Sticozzi,[1] Velio Bocci,[4] and Valter Travagli[3]

[1]Dipartimento di Biologia ed Evoluzione, Università degli Studi di Ferrara, Italy. [2]Department of Food and Nutrition, Kyung Hee University, Seoul, South Korea. [3]Dipartimento Farmaco Chimico Tecnologico, [4]Dipartimento di Fisiologia, Università degli Studi di Siena, Italy

Address for correspondence: Valter Travagli, Dipartimento Farmaco Chimico Tecnologico, Università degli Studi di Siena, Via Aldo Moro 2-53100 Siena, Italy. Phone: +39 0577 234317. Fax: +39 0577 234333. valter.travagli@unisi.it

The intervention strategies in various types of skin wounds include several treatment programs that depend on the identified disease. Several factors such as aging, defective nutrition, traumatism, atherosclerosis, and diabetes may contribute to the formation of a wound that has no tendency to heal due to a defective and complicated repair process. The numerous advances in the understanding of the wound-healing process in both acute and chronic lesions have been recently described. The purpose of this paper is to describe relatively new approaches as viable alternatives to current wound-healing therapies. The future challenges for both the best targeting and optimization of these potential treatments are also described.

Keywords: wound healing; oxidative stress; ROS; 1,2,4-oxolane moieties

Introduction

About 1.5 billion people suffer from skin diseases as a consequence of both progressive aging and the lack of adequate health-care. Among these, skin lesions are of great interest, and they can be mainly divided into acute or chronic wounds. Experts debate about the time for closure that defines a chronic nonhealing wound.[1–3]

It has been stated that "acute wounds generally follow trauma or inflammation and usually heal within six weeks."[4] Chronic wounds (in addition to failing to heal after six weeks) have characteristic pathological associations that inhibit or delay the healing process.[5] In particular, in the industrialized world it is estimated that 1–1.5% of the population undergoes problems related to recovering proper skin function, with particular reference to elderly and diabetic patients, or those with arteriosclerosis, who can easily suffer from ulcers and bedsores. The International Diabetes Federation estimates that 285 million people worldwide have diabetes.[6] This total is expected to rise to 438 million within 20 years. Moreover, type II diabetes, most often in connection with obesity, further increases the number of people with chronic wounds.

Throughout the world, beside keeping people from going to work, total expenses spent on the impaired healing of chronic wounds is enormous, about 2–4% of health spending, particularly because about 15% of type I diabetes patients must undergo amputations of the lower extremities. Certainly no less important are the anguish and suffering of the affected patients. Moreover, a number of issues directly related to treatment modalities and their correct evaluation are still open[7] despite the numerous recent advances in the understanding of the wound-healing process in both acute and chronic lesions.[8,9] Furthermore, since a wound contributes to a violation of natural defense barriers, infectious complications are very common, but should be readily avertible.[10] In this sense, the parenteral and/or topical use of valid antibiotics, antivirals, vaccines, and antiparasitic drugs makes a valuable contribution. Unfortunately, with time, pathogen agents may become drug resistant, and conventional drugs' side effects frequently limit their effectiveness. Another drawback is the cost, which compromises their use or their availability in poor countries. Moreover, wound patients are suffering not only because the drugs may become ineffective, but because they are discouraged by the slow healing process.

doi: 10.1111/j.1749-6632.2012.06636.x

Cutaneous wound healing is an age-related and sex-multiphase process,[11,12] and both innate and adoptive immune systems are too often hindered by the chronic infection. This is also the reason that could explain the failure of growth factors in heavily contaminated ulcers. This paper briefly summarizes the sequential phases of physiological wound healing and then analyses the causes that complicate the healing in chronic wounds. Moreover, after describing the pros and cons of current therapies, we fully describe both a relatively new approach, which has not been mentioned in recent relevant reviews,[13,14] as well as future concepts.

Physiological aspects of the wound repair process

In humans, and more widely in all mammalian species, the wound-healing process can be subdivided into three consecutive and overlapping phases: inflammation, tissue formation, and matrix formation and remodeling.[15] The transition from one phase to another depends on the maturation and differentiation of the main cell populations involved, among which keratinocytes, fibroblasts, neutrophils, and macrophages play the main roles.[15–18]

Recent observations show that stem cells have an unclear but likely major role in response to cutaneous injury,[19] as well as the evidence for the roles of M1 and M2 macrophage, and that of T cells.[20,21] The first event occurring after injury is the formation of a blood clot; several cells are involved in the blood plug: platelets, and red and white blood cells. With the action of fibrin fibers, the clot is stabilized and then "invaded" by several infiltrating cells, such as neutrophils, macrophages, mastocytes, platelets, and, possibly, by bacteria and toxins, which are counteracted by host-generated H_2O_2. Neutrophils massively infiltrating the wound during the first 24 hours postinjury are attracted by numerous inflammatory cytokines produced by activated platelets, endothelial cells, as well as by degradation products from pathogens. Macrophages massively infiltrating the wound two days postinjury produce intense phagocytic activity.[22]

After two to three days, the second phase lasts about two weeks and is characterized by neo-angiogenesis and granulation. During the re-epithelialization process, keratinocytes from the wound edges migrate over the wound bed at the interface between the wound dermis and the fibrin clot. This migration is facilitated by the production of specific proteases, such as collagenase by the epidermal cells to degrade the extracellular matrix. Activated fibroblasts also migrate to the wound bed and form, with the macrophages, granulation tissue. Intense angiogenesis, allowing the supply of oxygen and nutrients necessary for the healing process, also occurs within the tissue. Both growth factors and reactive oxygen species (ROS) produced by the granulation tissue will favor proliferation and differentiation of epithelial cells, restoring epithelial barrier integrity. The last stage of the wound-healing process consists in a gradual involution of the granulation tissue and dermal regeneration. This step is associated with apoptosis of myofibroblasts, endothelial cells, and macrophages. The remaining tissue is therefore composed mostly of extracellular matrix proteins, essentially collagen type III that will be remodeled by metalloproteinase produced by epidermal cells, endothelial cells, fibroblasts, and the macrophages remaining in the scar and then replaced by collagen type I.[8]

Chronic wound healing is mainly sustained by chronic inflammation, which without appropriate therapy tends to worsen. The basic reasons are not necessarily old age but rather hypertension and atherosclerosis, which can lead to ischemia, diabetes, and venous insufficiency. Common pathogenetic causes are local tissue hypoxia, edema, abundant bacterial colonization, and, possibly, repeated ischemia-reperfusion injuries. The surface area of a nonhealing wound tends to widen and shows fibrin deposition, necrotic areas, and a few islands of granulation tissue.

Current therapies

It is clear that delayed wound healing is due to the deficiency of essential parameters, such as a normal vascular bed, a physiological pO2, an active local immune system, and the normal sequential release of growth factors able to, step by step, heal the wound.[23] A number of therapies for chronic wounds have been evaluated, and they vary in relation to the causes of the wound, be they diabetes, peripheral arterial obstructive disease (PAOD), trauma, or venous insufficiency.[24,25] Moreover, infection complications and severity may vary from mild, when infection is limited to the skin surrounding the wound, to moderate, when infection has spread to nearby tissues, and to severe, with systemic toxicity, high fever, and metabolic problems. For antibiotic therapy to be effective, repeated culture data indicating the optimal antibiotics should be continued for at least two to three weeks.[26] However, such a subject cannot be considered as an emergent topic, except for the use of topical doxycycline to enhance healing of chronic wounds, because of its enzyme-inhibitor properties.[27] On the other hand, both the type of culture and the role of contamination have been long debated, together with the role of biofilm bacteria.[28,29] Several approaches useful for wound repair will be reviewed below.

Wound cleansing and debridement

Particular attention has been placed on measures of wound cleansing, as part of wound bed preparation, that gently and continuously removes debris and exudate to prepare the wound bed for closure.[30,31] It is increasingly well recognized that clearing a wound bed of nonviable tissue is an important step that may facilitate the healing process for a variety of wound types. In fact, removal of necrotic tissue is essential for the treatment of chronic wounds because devitalized tissue poses a major impediment to wound healing. A number of different modalities exist for wound debridement. The four most common debridement methods are surgical, autolytic, enzymatic, and mechanical.[32]

Wound care[33]

In order to prevent tissue dehydration and cell death and to enhance neo-angiogenesis and reepithelialization, the use of a clean, moist wound-healing environment appears to be very useful.[34] Recent advancements in technology and in the understanding of human physiology have led to the commercial development of dressings that offer material improvements with regard to these same ancient fundamental principles.[35]

Revascularization

Whenever possible, the treatment of peripheral arterial obstructive disease (PAOD) consists of performing endovascular or open surgery using either angioplasty or endarterectomy. Catheter-based options for revascularization currently play an important role. Available technology will continue to improve, and innovation will be rapid. These innovations include biodegradable stents, drug-eluting balloons, and new stent platforms.[36–38] Moreover, pharmaceutical therapy with statins and platelet antiaggregants, together with a low-fat diet and daily physical exercise are helpful.

Pressure offloading

The evidence for the role of offloading in the prevention and treatment of plantar ulcers in the diabetic foot has been assessed, even if it has been pointed out that there is a gap between evidence-based guidelines and current practice.[39] A highly evaluated offloading technique is represented by a contact casting, half shoe, and felted foam dressings to be changed every one to two weeks. Soft-tissue infections and osteomyelitis are contraindications to total-contact casting. Pressure offloading with the total-contact cast is the gold standard of care for chronic neuropathic noninfected, nonischemic plantar foot ulcers in individuals with diabetes mellitus. The recent trials of removable cast walkers rendered irremovable suggest that this alternate approach may be preferable, given that less technical expertise for fitting is required.[40]

Negative-pressure wound therapy

Negative-pressure wound therapy (NPWT) is a fairly popular method for providing subatmospheric pressure (about -125 mmHg) to dressed wounds connected to a vacuum pump, the area being sealed with adhesive film. There is little doubt that the negative pressure reduces the perilesional edema and hypoxia and stimulates cellular proliferation. This procedure has been found effective and safe for treating advanced diabetic foot ulcers,[41,42] notwithstanding the recent FDA Preliminary Public Health Notification and Advice for patients on serious complications, especially bleeding and infection, from the use of NPWT systems.[43] Although rare, these complications can occur wherever NPWT systems are used, including hospitals, long-term health-care facilities, or at home.[44]

Wound management agents

Recently, many agents have been used in the management of wounds. In detail, becaplermin, the recombinant human platelet-derived growth factor, subunit B, has been shown to be a potent mitogen for cells of mesenchymal origin. Binding of this growth factor to its receptor elicits a variety of cellular responses. It is released by platelets and plays an important role in stimulating adjacent cells to grow and thereby heal the wound.[45–47] For completeness sake, results of clinical trials conducted in developed Western countries cannot be directly extrapolated and applied to populations living in developing countries. In fact, a prospective comparative study, performed on 613 patients with diabetic foot lesions, documented the regional differences in risk factors and clinical presentations of diabetic foot lesions.[48]

Finally, other types of treatment options, such as the potential use of stem cells, have elicited interest, although, if stem cells are not autologous, they can be rejected.[49] Autologous cells may be used in some patients but not in diabetics because the cells can be defective. Gene therapy programmed in cells able to release growth factors is promising, but again release of growth factor in infected wounds is not effective.[50] Moreover, there is little evidence to justify the routine use of bioengineered skin substitutes, extracellular matrix protein surfaces, or dressing compared with standard care.[51,52]

A plethora of natural products and derivatives, such as terpenoids, have been more or less anecdotically used as wound-healing agents.[53,54] Among these, alkannins and shikonins—chiral pairs of naturally occurring isohexenylnaphthazarins found in the external layer of the roots of at least 150 species of the Boraginaceae family—show significant and promising wound-healing properties.[55]

As for nonnatural compounds as wound-healing agents, benzazepines, phenytoin, raxofelast, molsidomine, and *S*-nitrosothiols, together with nitric oxide in its gaseous form, represent leading compounds for the design of new efficient drugs.[54]

Proteolytic enzymes

The properties and functions of the extracellular and cell surface proteases involved in wound healing are well known. They involve matrix protein synthesis and degradation, release of cryptic bioactive fragments, regulation of growth factors and other cytokines, and the control of cell adhesion, migration, apoptosis, and signaling.[56] Moreover, it has been confirmed that collagenase ointment is a safe and effective choice for debridement of cutaneous ulcers and burn wounds.[57]

Maggot therapy

Maggot therapy is the application of disinfected fly larvae to chronic wounds. Maggots are gradually finding their way into a more acceptable system of wound management, and they have been used for treating diabetic foot ulcers unresponsive to conventional therapy.[58,59] Maggot secretions and excretions possess antibacterial activity against a wide range of pathogens and in their wound-healing capabilities in biosurgery.[60]

Hyperbaric oxygen therapy

Hyperbaric oxygen therapy (HBOT) delivers 100% O_2 at 2–3 bar for a period of one to two hours usually five days a week in a suitable chamber. The leading idea is to enhance the low pO_2 values of hypoxic tissues and to display some bactericidal effect. Normally, 10–30 treatments are necessary if the patient tolerates them. On the whole, HBOT improves the chance of healing and reduces the risk of amputation of the lower extremities.[61–63] However, even if there is practically no risk of chamber explosion, it is a costly therapy;[64] both tympanic membrane rupture and pneumothorax are common side effects,[65] and presently, there are conflicting data regarding the efficacy of this therapy.[63,66,67] It may be given in conjunction with topical oxygen therapy localized to the wound area using a portable inflatable device. This method lowers the risk of oxygen toxicity, improves the oxygenation of the lesion, has a low cost, and has the advantage of the option of home treatment. Kalliainen *et al.* evaluated this topical therapy and found that it was beneficial in improving the wound.[68]

We have described the current main types of treatments for cutaneous wound healing; there are also emerging topics in cutaneous wound repair involving the use of substances correlated with ROS, with particular emphasis on the use of ozone and its derivatives, relating to cellular and humoral responses to cutaneous injuries. Such topics will be discussed below.

Role of reactive oxygen species in wound repair

As mentioned before, the wound-healing process is regulated by a large variety of different growth factors, such as cytokines and hormones,[69] but also by ROS. Such factors, among which the superoxide anion ($^-O_2$) is central because it may be converted into other physiologically relevant ROS by enzymatic or nonenzymatic reactions, are required for the defense against invading pathogens,[70] and low levels of ROS are also essential mediators of intracellular signaling.[71] It has been shown that low doses of H_2O_2 can improve wound healing. On the other hand, excessive amounts of ROS are deleterious due to their high reactivity.[72,73] Because of the short half-life of ROS, their concentrations *in vivo* are difficult to determine, although the levels of H_2O_2 have been recently measured in wound fluid from acute murine excisional wounds.[74] These studies revealed that concentrations of H_2O_2, ranging from 100 μM to 250 μM, are present at the wound site and these levels are even higher in the first stage of wound healing.

It has, therefore, been suggested that the low levels of ROS that are produced in normal wounds are important for the repair process. In other studies the wound levels of ROS have been determined indirectly through analysis of oxidation products of lipids, proteins, or DNA.[75] A major product of lipid peroxidation is 4-hydroxy-2-nonenal (4-HNE), which could be detected by immunohistochemistry at the edge of murine excisional wounds. Interestingly, coimmunostaining revealed that 4-HNE mainly colocalizes with neutrophils, suggesting that the respiratory burst of these inflammatory cells results in the production of superoxide, which in turn causes lipid peroxidation.[76]

Topical use of ozone and its derivatives as mediators of ROS

The use of ozone is a neglected but important method for the treatment of chronic wounds. Since its use is increasing in the scientific community, and to stimulate interest about this topic, we discuss the use of ozone here in more detail.

It is generally understood that, although O_3 is not a radical species per se, the toxic effects of O_3 are mediated through free radical reactions; they are achieved either directly by the oxidation of biomolecules to produce the classical radical species (hydroxyl radical), H_2O_2, or by driving the radical-dependent production of cytotoxic, nonradical species (aldehydes).[77,78] The first mention of the use of ozone for treating dermatological pathologies dates back to the nineteenth century, when it was identified as a potent anti-infective gas; ozone was used during World War I for treating German soldiers affected by gaseous gangrene due to *Clostridium* anaerobic infections.[79] Recently, in Germany, Italy, Russia, and Cuba, the possibility of using ozone in its various forms as an anti-infective in veterinary and human medicine has been evaluated.[80,81] In fact, ozone is slowly being used in several forms, as a gas, as ozonated water, or as ozonated natural matrices in a variety of infections, trophic ulcers, burns, cellulitis, abscesses, anal fissures, decubitus in paralytic patients, fungal diseases, gingivitis, peritonitis, and vulvovaginitis.[82,83] It is generally understood that ozone, under various formulations, can display a cleansing effect and, due to its oxidative properties, act as a potent disinfectant against many pathogens present in the skin and mucosal surfaces. However, the bactericidal action of ozone, rapidly effective in contaminated water, is markedly reduced when antioxidant compounds are present.

Our studies, performed with various bacterial suspensions either in pure saline medium or saline in addition to human serum, have in fact yielded discouraging results: the presence of only 5–10% of serum blocks much of ozone's bactericidal effect. Such an outcome is not surprising because serum shows a potent antioxidant capacity able to neutralize the ozone oxidant effect.[84] On the other hand, the above data appear to be instructive in the sense that the application of ozonated derivatives, mainly ozonated oils, should—if the ozone is to be effective—be preceded by careful cleaning and washing of the biological exudates present in wounds and ulcers. It seems also useful to have a series of ozonated oils that have gradations of peroxide value used on wounds that are highly contaminated with bacteria mixed with dead cells. On the other hand, the decomposition of ozone derivatives has the additional advantage of improving the local metabolism and proliferation of tissues essential for eventual mucosal or/and cutaneous healings.[85]

Today, in modestly developed countries especially, the utility of ozone is greatly valued, while in highly technologically advanced countries ozone remains underused because of prejudice, lack of information about its utility, and the wide availability of pharmaceutical products, which are not always effective.

The use of topical ozone can be applied in two ways: (a) the bagging of the ulcerated lesion with exposure to a gaseous oxygen/ozone mixture for

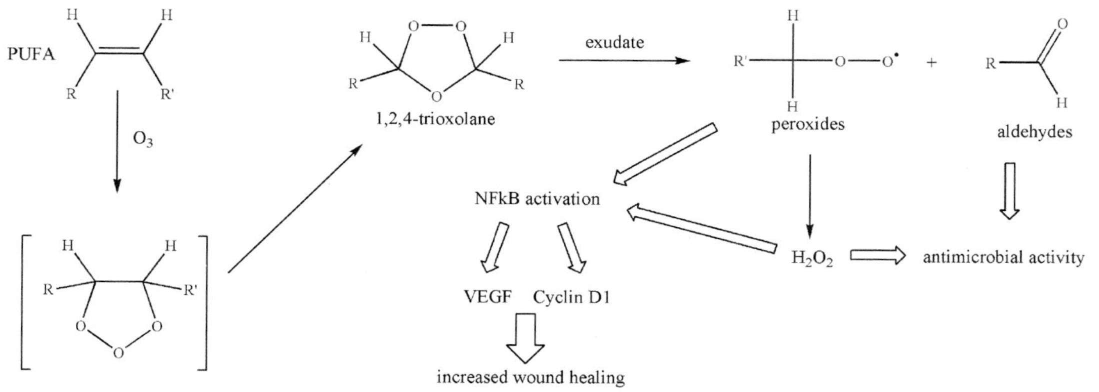

Figure 1. Schematic of an ozonated derivative formation by chemical reaction of ozone with unsaturated fatty acids of triglycerides from oils, and their possible mechanisms. The interaction between ozone and PUFA leads to the formation of aldehydes, peroxides, and H_2O_2 that can affect wound-healing process by activating redox-sensitive transcription factors (NF-κB) that are responsible for the expression of proangiogenic (VEGF) and proliferative (Cyclin D1) genes involved in the wound-healing processes.

about 30 minutes daily, using an ozone concentration of 80 μg/mL down to 5–10 μg/mL in clean wounds; or (b) the local treatment of the wound with ozonated water or ozonated saline, again using variable concentrations of ozone. This is considered a chemically effective debridement, where pus, deposits of fibrin, and necrotic tissue are removed with the consequent activation and oxygenation of the wound. This should be followed by the application of ozonated oil, which keeps the wound sterile and activated through the release of ozonated products and oxygen. In addition, ozonated oils with reliable peroxide values can also be used for wound-healing purposes.[85] By what mechanism ozonated oils act remains an open question (Fig. 1). Most likely, when the stable triozonide comes into contact with the warm exudate of the wound, the triozonide slowly decomposes into different peroxides, which readily dissolves in water, probably generating hydrogen peroxide, which can explain the prolonged disinfectant and stimulatory activity. Consequently, titrated preparations should be used with high, medium, or low ozonide concentrations during the inflammatory septic phase I, the regenerating phase II, or the remodeling phase III, respectively. These phases have been related to the rapidly changing cell types and to the release of cytokines and growth factors that modulate the complex healing process.

Future challenges

In recent years, research in the pharmaceutical field has turned toward a growing commitment in the development of new technologies to optimize applications and increase bioavailability, while minimizing potential side effects of substances applied to both skin and mucosae. In that regard, a great interest in new techniques that enhance the permeation of drugs at the different membrane levels has been developed. In particular, the topical application of drugs through lipid colloidal carriers, such as vesicular systems (traditional liposomes and ethosomes) and solid lipid nanoparticles (SLN), has stirred much interest.[86,87] These carriers are proposed for dermal and topical application of useful substances and have a number of advantages: they enhance the penetration of both lipophilic and hydrophilic substances by incorporation; they have high affinity and similarity with the epidermal barrier, offering the possibility of improving the absorption of drugs across the skin barrier, thus ensuring a greater local concentration; they are natural (bio-compatible, nontoxic, nonimmunogenic); and they have a natural skin-moisture action because their small size gives them adhesive properties to form a lipid film on the skin surface. Topical application allows for reconstruction of the lipid layer and alters the skin barrier damaged by various diseases. Eventually, we will have greater control of the rate of supply of therapeutic agents. As it stands now, either the carrier-to-skin or the carrier-to-mucosa contact medium is absorbed from the surface and slowly releases its biologically active content.

Moreover, of great interest is the potential application of micro-and nanobubbles for the targeted release of efficient drugs, including ozone and its derivatives. Bubbles may be used in a number of ways to aid drug delivery; for example, drugs may be encapsulated within the bubbles—incorporated into the shell or, for example, by ligands embedded in the lipid membrane.[88,89] It may also be possible to construct microbubbles with a multilayered shell containing a given drug. The spatial localization would rely on the ability to confine the ultrasound beam to the required volume.[90] If such bubbles can be accumulated within the target volume, ultrasound can destroy them locally, releasing the therapeutic agent able to influence neuronal, stromal, vascular, and circulatory system cells by chemical, physical, or biological stimuli.[91]

Conflicts of interest

The authors declare no conflicts of interest.

References

1. Dealey, C. 2005. *The Care of Wounds: A Guide for Nurses*. 3rd ed. Blackwell Publishing Ltd. Oxford, UK.
2. Whitney, J.D. 2005. Overview: acute and chronic wounds. *Nurs. Clin. North Am.* **40:** 191–205.
3. Bryant, R.A. & D. Nix. 2011. *Acute and Chronic Wounds, Current Management Concepts*. 4th ed. Mosby Elsevier. Missouri, USA.
4. Kumar, S. & D.J. Leaper. 2008. Classification and management of acute wounds. *Surgery* **26:** 43–47.
5. Jones, K.R., K. Fennie & A. Lenihan. 2007. Evidence-based management of chronic wounds. *Adv. Skin Wound Care* **20:** 591–600.
6. Understand diabetes, take control. Available at http://www.idf.org/worlddiabetesday/2009-2013/booklet (last accessed May 15, 2012).
7. Gottrup, F. & T. Karlsmark. 2009. Current management of wound healing. *G. Ital. Dermatol. Venereol.* **144:** 217–228.
8. Singer, A.J. & R.A. Clark. 1999. Cutaneous wound healing. *N. Engl. J. Med.* **341:** 738–746.
9. Singer, A.J. & A.B. Dagum. 2008. Current management of acute cutaneous wounds. *N. Engl. J. Med.* **359:** 1037–1046.
10. Bowler, P.G., B.I. Duerden & D.G. Armstrong. 2001. Wound microbiology and associated approaches to wound management. *Clin. Microbiol. Rev.* **14:** 244–269.
11. Swift, M.E., A.L. Burns, K.L. Gray & L.A. Di Pietro. 2001. Age-related alterations in the inflammatory response to dermal injury. *J. Invest. Dermatol.* **117:** 1027–1035.
12. Gilliver, S.C., J.J. Ashworth, S.J. Mills, M.J. Hardman & G.S. Ashcroft. 2006. Androgens modulate the inflammatory response during acute wound healing. *J. Cell Sci.* **119:** 722–732.
13. Schreml, S., R.M. Szeimies, L. Prantl, S. Karrer, M. Landthaler & P. Babilas. 2010. Oxygen in acute and chronic wound healing. *Br. J. Dermatol.* **163:** 257–268.
14. Tecilazich, F., T. Dinh & A. Veves. 2011. Treating diabetic ulcers. *Expert Opin. Pharmacother.* **12:** 593–606.
15. Rodero, M.P. & K. Khosrotehrani. 2010. Skin wound healing modulation by macrophages. *Int. J. Clin. Exp. Pathol.* **3:** 643–653.
16. Leibovich, S.J. & R. Ross. 1975. The role of the macrophage in wound repair. A study with hydrocortisone and antimacrophage serum. *Am. J. Pathol.* **78:** 71–100.
17. Deonarine, K., M.C. Panelli, M.E. Stashower, *et al.* 2007. Gene expression profiling of cutaneous wound healing. *J. Transl. Med.* **5:** 11.
18. Becker, D.L., C. Thrasivoulou, A.R. Phillips. 2011. Connexins in wound healing; perspectives in diabetic patients. *Biochim. Biophys Acta* **1818:** 2068–2075.
19. Lanza R. 2004. *Handbook of Stem Cells*. Academic Press. Boston, MA.
20. Gilliver, S.C., E. Emmerson, J. Bernhagen & M.J. Hardman. 2011. MIF: a key player in cutaneous biology and wound healing. *Exp. Dermatol.* **20:** 1–6.
21. Sindrilaru, A., T. Peters, S. Wieschalka, *et al.* 2011. An unrestrained proinflammatory M1 macrophage population induced by iron impairs wound healing in humans and mice. *J. Clin. Invest.* **121:** 985–997. doi: 10.1172/JCI44490. Epub 2011 Feb 7.
22. Mosser, D.M. & J.P. Edwards. 2008. Exploring the full spectrum of macrophage activation. *Nature Rev. Immunol.* **8:** 958–969.
23. Braiman-Wiksman, L., I. Solomonik, R. Spira & T. Tennenbaum. 2007. Novel insights into wound healing sequence of events. *Toxicol. Pathol.* **35:** 767–779.
24. Grey J.E. & G.K. Harding. 2006. *ABC of Wound Healing*. Wiley-Blackwell. New York.
25. Springer, M.L. 2006. A balancing act: therapeutic approaches for the modulation of angiogenesis. *Curr. Opin. Investig. Drugs* **7:** 243–250.
26. Eaglstein, W.H., P.M. Mertz & O.M. Alvarez. 1984. Effect of topically applied agents on healing wounds. *Clin. Dermatol.* **2:** 112–115.
27. Stechmiller, J., L. Cowan & G. Schultz. 2010. The role of doxycycline as a matrix metalloproteinase inhibitor for the treatment of chronic wounds. *Biol. Res. Nurs.* **11:** 336–344.
28. James, G.A., E. Swogger, R. Wolcott, *et al.* 2008. Biofilms in chronic wounds. *Wound Repair Regen.* **16:** 37–44.
29. Fazli, M., T. Bjarnsholt, K. Kirketerp-Møller, *et al.* 2011. Quantitative analysis of the cellular inflammatory response against biofilm bacteria in chronic wounds. *Wound Repair Regen.* **19:** 387–391.
30. Andriessen, A.E. 2010. Assessment of a wound cleansing solution in the treatment of problem wounds. *Wounds* **20:** 171–175.
31. National Pressure Ulcer Advisory Panel & European Pressure Ulcer Advisory Panel. 2009. Pressure ulcer treatment recommendations. In: *Prevention and Treatment of Pressure Ulcers: Clinical Practice Guideline*. National Pressure Ulcer Advisory Panel. Washington (DC). pp. 51–120.
32. Shi, L., R. Ermis, K. Lam, *et al.* 2009. Study on the debridement efficacy of formulated enzymatic wound debriding agents by in vitro assessment using artificial wound eschar and by an in vivo pig model. *Wound Repair Regen.* **17:** 853–862.

33. Sussman C & B. Bates-Jensen. 2007 *Wound Care: A Collaborative Practice Manual.* Lippincott Williams & Wilkins. Baltimore.
34. Cowan, L.J. & J. Stechmiller. 2009. Prevalence of wet-to-dry dressings in wound care. *Adv. Skin Wound Care* **22:** 567–573.
35. Lee, J.C., S. Kandula & N.S. Sherber. 2009. Beyond wet-to-dry: a rational approach to treating chronic wounds. *Eplasty* **9:** 131–137.
36. Tautenhahn, J., R. Lobmann, B. Koenig, *et al.* 2008. The influence of polymorbidity, revascularization, and wound therapy on the healing of arterial ulceration. *Vasc. Health Risk Manag.* **4:** 683–689.
37. Kirksey, L. & M. Troiano. 2011. The evolving paradigm of revascularization for wound healing: which intervention for which wound? *Wounds* **23:** 49–52.
38. Neville, R.F. 2011. Open surgical revascularization for wound healing: Past performance and future directions. *Plast. Reconstr. Surg.* **127:** 154S–162S.
39. Cavanagh, P.R. & S.A. Bus. 2010. Off-loading the diabetic foot for ulcer prevention and healing. *J. Vasc. Surg.* **52:** 37S–43S.
40. Hunt, D.L. 2011. Diabetes: foot ulcers and amputations. *Clin. Evid. (Online).* **2011**: pii 0602.
41. Armstrong, D.G., L.A. Lavery & Diabetic Foot Study Consortium. 2005. Negative pressure wound therapy after partial diabetic foot amputation: a multicentre, randomised controlled trial. *Lancet* **366:** 1704–1710.
42. Xie, X., M. McGregor & N. Dendukuri. 2010. The clinical effectiveness of negative pressure wound therapy: a systematic review. *J. Wound Care* **19:** 490–495.
43. FDA safety communication: UPDATE on serious complications associated with negative pressure wound therapy systems. Date Issued February 24, 2011. Available at http://www.fda.gov/MedicalDevices/Safety/AlertsandNotices/ucm244211.htm (last accessed May 15, 2012)
44. Gregor, S., M. Maegele, S. Sauerland, *et al.* 2008. Negative pressure wound therapy. A vacuum of evidence? *Arch. Surg.* **143:** 189–196.
45. Smiell, J.M., T.J. Wieman, D.L. Steed, *et al.* 1999. Efficacy and safety of becaplermin (recombinant human platelet-derived growth factor-BB) in patients with nonhealing, lower extremity diabetic ulcers: a combined analysis of four randomized studies. *Wound Repair Regen.* **7:** 335–346.
46. Papanas, N. & E. Maltezos. 2010. Benefit-risk assessment of becaplermin in the treatment of diabetic foot ulcers. *Drug Safety* **33:** 455–461.
47. Buchberger, B., M. Follmann, D. Freyer, *et al.* 2011. The evidence for the use of growth factors and active skin substitutes for the treatment of non-infected diabetic foot ulcers (DFU): a health technology assessment (HTA). *J. Exp. Clin. Endocrinol. Diabetes* **119:** 472–479.
48. Hardikar, J.V., Y. Chiranjeev Reddy, D.D. Bung, *et al.* 2005. Efficacy of recombinant human platelet-derived growth factor (rhPDGF) based gel in diabetic foot ulcers: a randomized, multicentre, double-blind, placebo-controlled study in India. *Wounds* **17:** 14–152.
49. Cha, J. & V. Falanga. 2007. Stem cells in cutaneous wound healing. *Clin. Dermatol.* **25:** 73–78.
50. Hirsch, T., M. Spielmann, F. Yao & E. Eriksson. 2007. Gene therapy in cutaneous wound healing. *Front. Biosci.* **12:** 2507–2518.
51. Reddy, M., S.S. Gill, S.R. Kalkar, *et al.* 2008. Treatment of pressure ulcers: a systematic review. *JAMA* **300:** 2647–2662.
52. Dumville, J.C., S. O'Meara, S. Deshpande & K. Speak. 2011. Hydrogel dressings for healing diabetic foot ulcers. *Cochrane Database Syst. Rev.* **9:** CD009101.
53. Kumar, B., M. Vijayakumar, R. Govindarajan & P. Pushpangadan. 2007. Ethnopharmacological approaches to wound healing-exploring medicinal plants of India. *J. Ethnopharmacol.* **114:** 103–113.
54. de Fátima, A., L.V. Modolo, A.C. Sanches & R.R. Porto. 2008. Wound healing agents: the role of natural and non-natural products in drug development. *Mini Rev. Med. Chem.* **8:** 879–888.
55. Papageorgiou, V.P., A.N. Assimopoulou & A.C. Ballis. 2008. Alkannins and shikonins: a new class of wound healing agents. *Curr. Med. Chem.* **15:** 3248–3267.
56. Moali, C. & D.J. Hulmes. 2009. Extracellular and cell surface proteases in wound healing: new players are still emerging. *Eur. J. Dermatol.* **19:** 552–564.
57. Ramundo, J. & M. Gray. 2009. Collagenase for enzymatic debridement: a systematic review. *J. Wound Ostomy Continence Nurs.* **36:** S4–S11.
58. Sherman, R.A. 2003. Maggot therapy for treating diabetic foot ulcers unresponsive to conventional therapy. *Diabetes Care* **26:** 446–451.
59. Gupta, A. 2008. A review of the use of maggots in wound therapy. *Ann. Plast. Surg.* **60:** 224–227.
60. Arora, S., C. Baptista & C.S. Lim. 2011. Maggot metabolites and their combinatory effects with antibiotic on Staphylococcus aureus. *Ann. Clin. Microbiol. Antimicrob.* **10:** 6.
61. Smith, B.M., L.D. Desvigne, J.B. Slade, *et al.* 1996. Transcutaneous oxygen measurements predict healing of leg wounds with hyperbaric therapy. *Wound Repair Regen.* **4:** 224–229.
62. Fife, C.E., C. Buyukcakir, G.H. Otto, *et al.* 2002. The predictive value of transcutaneous oxygen tension measurement in diabetic lower extremity ulcers treated with hyperbaric oxygen therapy: a retrospective analysis of 1,144 patients. *Wound Repair Regen.* **10:** 198–207.
63. Kranke, P., M. Bennett, I. Roeckl-Wiedmann & S. Debus. 2004. Hyperbaric oxygen therapy for chronic wounds. *Cochrane Database Syst. Rev.* **2:** CD004123.
64. Flegg, J.A., D.L. McElwain, H.M. Byrne & I.W. Turner. 2009. A three species model to simulate application of Hyperbaric Oxygen Therapy to chronic wounds. *PLoS Comput. Biol.* **5:** e1000451.
65. Kaur, S., M. Pawar, N. Banerjee & R. Garg. 2012. Evaluation of the efficacy of hyperbaric oxygen therapy in the management of chronic nonhealing ulcer and role of periwound transcutaneous oximetry as a predictor of wound healing response: a randomized prospective controlled trial. *J. Anaesthesiol. Clin. Pharmacol.* **28:** 70–75.
66. Kranke, P., M. Bennett, I. Roeckl-Wiedmann & S. Debus. 2004. Hyperbaric oxygen therapy for chronic wounds. *Cochrane Database Syst. Rev.* **4:** CD004123.

67. Berendt, A.R. 2006. Counterpoint: hyperbaric oxygen for diabetic foot wounds is not effective. *Clin. Infect. Dis.* **43:** 193–198.
68. Kalliainen, L.K., G.M. Gordillo, R. Schlanger & C.K. Sen. 2003. Topical oxygen as an adjunct to wound healing: a clinical case series. *Pathophysiology* **9:** 81–87.
69. Werner, S. & R. Grose. 2003. Regulation of wound healing by growth factors and cytokines. *Physiol. Rev.* **83:** 835–870.
70. Arul, V., J.G. Masilamoni, E.P. Jesudason, *et al.* 2012. Glucose oxidase incorporated collagen matrices for dermal wound repair in diabetic rat models: a biochemical study. *J. Biomater. Appl.* **26:** 917–938.
71. Valacchi, G., V. Fortino & V. Bocci. 2005. The dual action of ozone on the skin. *Br. J. Dermatol.* **153:** 1096–1100.
72. auf dem Keller, U., A. Kümin, S. Braun & S. Werner. 2006. Reactive oxygen species and their detoxification in healing skin wounds. *J. Investig. Dermatol. Symp. Proc.* **11:** 106–111.
73. Vermeij, W.P. & C. Backendorf. 2010. Skin cornification proteins provide global link between ROS detoxification and cell migration during wound healing. *PLoS One.* **5:** e11957.
74. Roy, S., S. Khanna, K. Nallu, T.K. Hunt & C.K. Sen. 2006. Dermal wound healing is subject to redox control. *Mol. Ther.* **13:** 211–220.
75. Schäfer, M. & S. Werner. 2008. Oxidative stress in normal and impaired wound repair. *Pharmacol. Res.* **58:** 165–171.
76. Ojha, N., S. Roy, G. He, *et al.* 2008. Assessment of wound-site redox environment and the significance of rac2 in cutaneous healing. *Free Radic. Biol. Med.* **44:** 682–691.
77. Pryor, W.A. 1994. Mechanisms of radical formation from reactions of ozone with target molecules in the lung. *Free Radic. Biol. Med.* **17:** 451–465.
78. Sathishkumar, K., M. Haque, T.E. Perumal, *et al.* 2005. A major ozonation product of cholesterol, 3β-hydroxy-5-oxo-5,6-secocholestan-6-al, induces apoptosis in H9c2 cardiomyoblasts. *FEBS Lett.* **579:** 6444–6450.
79. Stoker, G. 1916. The surgical uses of ozone. *Lancet* **188:** 712.
80. Bocci, V., E. Borrelli, V. Travagli & I. Zanardi. 2009. The ozone paradox: ozone is a strong oxidant as well as a medical drug. *Med. Res. Rev.* **29:** 646–682.
81. Bocci, V. 2011. *Ozone. A New Medical Drug*. Springer. Dordrecht, The Netherlands.
82. Travagli, V., I. Zanardi & V. Bocci. 2009. Topical applications of ozone and ozonated oils as anti-infective agents: an insight into the patent claims. *Recent Pat. Antiinfect. Drug Discov.* **4:** 130–142.
83. Travagli, V., I. Zanardi, G. Valacchi & V. Bocci. 2010. Ozone and ozonated oils in skin diseases: a review. *Mediators Inflamm.* **2010:** 610418.
84. Burgassi, S., I. Zanardi, V. Travagli, *et al.* 2009. How much ozone bactericidal activity is compromised by plasma components? *J. Appl. Microbiol.* **106:** 1715–1721.
85. Valacchi, G., Y. Lim, G. Belmonte, *et al.* 2011. Ozonated sesame oil enhances cutaneous wound healing in SKH1 mice. *Wound Repair Regen.* **19:** 107–115.
86. Cosco, D., C. Celia, F. Cilurzo, *et al.* 2008. Colloidal carriers for the enhanced delivery through the skin. *Expert Opin. Drug Deliv.* **5:** 737–755.
87. Jeschke, M.G., G. Sandmann, C.C. Finnerty, *et al.* 2005. The structure and composition of liposomes can affect skin regeneration, morphology and growth factor expression in acute wounds. *Gene Ther.* **12:** 1718–1724.
88. Cavalli, R., A. Bisazza, P. Giustetto, *et al.* 2009. Preparation and characterization of dextran nanobubbles for oxygen delivery. *Int. J. Pharm.* **381:** 160–165.
89. Yoon, C.S., H.S. Jung, M.J. Kwon, *et al.* 2009. Sonoporation of the minicircle-VEGF(165) for wound healing of diabetic mice. *Pharm. Res.* **26:** 794–801.
90. Johnson S. 2003. Low-frequency ultrasound to manage chronic venous leg ulcers. *Br. J. Nurs.* **12:** S14–S24.
91. ter Haar, G. 2007. Therapeutic applications of ultrasound. *Prog. Biophys. Mol. Biol.* **93:** 111–129.

Ann. N.Y. Acad. Sci. ISSN 0077-8923

ANNALS OF THE NEW YORK ACADEMY OF SCIENCES
Issue: *Environmental Stressors in Biology and Medicine*

Corrigendum for Ann. N.Y. Acad. Sci. 1236: 30–43

Marder, S.R., B. Roth, P.F. Sullivan, *et al.* 2011. Advancing drug discovery for schizophrenia. *Ann. N.Y. Acad. Sci.* 1236: 30–43.

In the above-cited article, Akira Sawa was inadvertently misspelled as Akira Saw.

The correct author list is:

Stephen R. Marder, Bryan Roth, Patrick F. Sullivan, Edward M. Scolnick, Eric J. Nestler, Mark A. Geyer, Daniel R. Welnberger, Maria Karayiorgou, Alessandro Guidotti, Jay Gingrich, Schahram Akbarian, Robert W. Buchanan, Jeffrey A. Lieberman, P. Jeffrey Conn, Stephen J. Haggarty, Amanda J. Law, Brian Campbell, John H. Krystal, Bita Moghaddam, Akira Sawa, Marc G. Caron, Susan R. George, John A. Allen, and Michelle Solis

doi: 10.1111/j.1749-6632.2012.06711.x